SpringerBriefs in Anthropology

Human Behavior, Biology and Evolution

Just a few years after the 150 anniversary of the publication of "Origins of the Species" and official Year of Darwin, the social and biological sciences are still working to integrate their research into a more complete understanding of human behavior and biology. This series hopes to close this gap as demonstrated by the potential topics and their audiences: anthropologists, biologists, demographers, economists, psychologists, geneticists, public health researchers, and others. This series will closely model the aims and scope of the journal Human Nature. It will focus on human evolutionary ecology, biology, psychology and behavior. The Briefs are based on evolutionary and life history theory. The social, behavioral, and biological sciences have traditionally pursued separate paths both in research and in education. Yet it is increasingly apparent that the most pervasive and crucial issues of our time are fundamentally behavioral in their nature and lie squarely at the interface of these artificially separated disciplines.

To address the problems of modern society, we must first understand the very substance of our species: how we evolved to be the way we are, why we behave the way we do, and what are the social and ecological contexts that restrain, channel, modify, and diversify our behavior. To this end, this Springer Brief series will look through a complex, multi-faceted lens: one that peers through time, across geographic space and into the diversity of human social and cultural experience. It will emphasize the integration of biological, environmental, social, psychological and behavioral factors that shape the expression of human behavior and the relevance of a biosocial perspective to major scientific, social, and policy issues in the world today.

More information about this subseries at http://www.springer.com/series/15166

Anne Campbell • Lee T Copping •
Catharine P Cross

Sex Differences in Fear Response

An Evolutionary Perspective

Springer

Anne Campbell (1951–2017)
Professor Anne Campbell wrote this book while at Durham University
Durham, UK

Lee T Copping
Teesside University
Middlesbrough, UK

Catharine P Cross
University of St Andrews
St Andrews, UK

ISSN 2195-0806 ISSN 2195-0814 (electronic)
SpringerBriefs in Anthropology
ISSN 2366-8792 ISSN 2366-8806 (electronic)
Human Behavior, Biology and Evolution
ISBN 978-3-030-65279-1 ISBN 978-3-030-65280-7 (eBook)
https://doi.org/10.1007/978-3-030-65280-7

This Springer imprint is published by the registered company Springer Nature Switzerland AG.
The registered company address is: Gewerbestrasse 11, 6330 Cham, Switzerland

Contents

Chapter 1
Fear, Sex Differences and the 'Staying Alive' Hypothesis

Fear is "an evolutionarily conserved affective motivational system that can be activated under conditions of novelty, sudden or intense stimulation, reactions to danger prepared by evolution, social interactions with unfamiliar conspecifics and conditioned responses to punishment" (Rothbart & Bates, 1998, p. 109). It is "a system that detects danger and produces responses that maximise the probability of surviving a dangerous situation" (LeDoux, 1998, p. 128). The adaptive benefits of an efficient fear system are therefore evident as the speedy recognition and avoidance of life-threatening dangers enhance survival and thus improve the chances of leaving offspring who also survive and reproduce. While intuitively it seems that the evolutionary advantages of staying alive should be the same for both sexes in mammals (such as humans), this might not be the whole story (Campbell, 1999, 2013). In this volume, we focus on why this might not be the case, and specifically why an adapted fear system may help explain sex differences in behaviours such as risk taking and aggression as a way of ensuring the survival of females and their offspring.

1.1 The Evolution of Sex Differences

Before we begin discussing the central premise of this review, it is necessary to briefly examine evolutionary theory surrounding why sex differences in behaviours may emerge to begin with. This requires a brief overview of two prominent evolutionary theories: sexual selection theory (Darwin, 1871/2004) and parental investment theory (Trivers, 1972).

Although both sexes are equally represented at a population level (due to the Fisher condition—Fisher, 1930), a scarcity of females impacts upon population growth more seriously than a lack of males. This is because females invest more heavily in offspring (in terms of time, energy and resources) and are more crucial to their survival in the early stages of their development. As females make a higher

A. Campbell et al., *Sex Differences in Fear Response*, SpringerBriefs in Anthropology, https://doi.org/10.1007/978-3-030-65280-7_1

parental investment in terms of gestation and lactation, they are excluded from the mating 'pool' for protracted periods of time (the group of individuals who are ready to mate). This creates a male-biased operational sex ratio so that fertile females are a limiting resource for whom males must compete with each other for access. Selection pressures thus act to increase males's greater sex drive, aggression, upper body strength and risk-taking (see Fromhage & Jennions, 2016) in order to increase the chances of their reproductive success. These attributes allow males to seize resources, deter and/or remove competition and monopolize access to fertile mates. These same attributes however also act as a guide for female mate choice, allowing the winners of these competitions to effectively showcase their high genetic quality. Sexual selection thus maintains sex differences within a population through intra-male conflict and female mate choice and is invoked as an explanation for many psychological and behavioural sex differences in both humans (Archer, 2019) and non-humans (Fromhage & Jennions, 2016).

Allied to this concept, Trivers (1972) argued that sex differences emerge due to differing parental investment strategies of males and females. The sex with the least investment in offspring (typically the male) is forced to compete for mate access. Conversely, the sex with the largest investment (females in the vast majority of species) aims to maximise fitness returns through careful selection of high-quality mates. These differences reflect *obligatory* levels of investment however, not necessarily *typical* levels. Differential investment strategies normally manifest in a continuum of low to high parental investment strategies in both males and females, contingent on local ecological conditions (such as assortative mating and mutual mate choice; Gangestad & Simpson, 2000; Jennions & Petrie, 1997; Kokko & Jennions, 2008; Simpson & Gangestad, 1991). The differential focus on mating versus parenting can thus lead to the maintenance of sex differences in behaviours within populations.

Males (having the smallest obligate investment) are prone to invest more time in mating effort, which often leads to increased intra-sexual competition between them. Those who succeed produce more offspring and pass their genes onto the next generation. The genes responsible for whatever gave them the successful competitive edge are propagated also. Over successive generations, this can result in the emergence of sexually dimorphic characteristics and behaviours. As such, sexual selection cogently explains why there are sex differences in behaviours aligned to achieving reproductive ends. Males must compete, which is an inherently risky business where loss can mean lineage extinction. Those that take risks have the opportunity to be very successful, while many will opt for less risky alternatives by investing more effort in parenting. This in turn creates variation in reproductive success (Betzig, 2012; Brown, Laland, & Borgerhoff Mulder, 2009). Risk taking, behaviors aligned to it (such as aggression) and other behaviours adapted to promote successful competition would therefore be expected to show sex differences favouring males (a point we will return to in detail in Chap. 2).

Historically, most behaviours exhibiting sex differences are examined through the lens of sexual selection, focusing on the key role of male-male competition (Wilson & Daly, 1985). This perspective however largely ignores the fact that females too aim to increase their own reproductive fitness, often at the expense of their competitors, and that their investment is critical to offspring survival (Campbell, 1999, 2009). There is debate over the narrowness of the term 'sexual selection' as envisioned by Darwin (1871/2004) who himself acknowledged that natural and sexual selection are not always clearly distinguishable (Kavanagh, 2006; Roughgarden, Oisha, & Akcay, 2006). Some have suggested that the term sexual selection should be more inclusive (Campbell, 2009) and others so far as to suggest that a new term, 'sex-dependent selection' should be ascribed to processes where the impact of natural selection forces would differ between males and females as a result of their different reproductive strategies (Carranza, 2009). In what follows, where the focus of our discussion will be on the benefits of risk aversion to women, when we apply the term 'selection' we are referring to selection forces that would meet this broader definition (unless stated otherwise).

1.2 Staying Alive

So far, we have suggested that the male propensity for risk taking emerged via sexual selection as a response to mating competition. In this section, we review an evolutionary scenario that suggests why selection may have had a parallel role in reducing the propensity for risk in women; by selecting for increased desistance from dangerous situations via a heightened sensitivity to fear (Campbell, 1999). There are good reasons to suggest why this would have been necessary as a survival function, particularly in modern hominids (on which this volume will centre its discussion).

Temperamentally, low levels of fear in response to danger enhance risk-taking and ultimately improve male reproductive success (Wilson & Daly, 1985). Competing for mates or fighting for resources often entails risky but necessary actions that are intended to result in fitness payoffs. High levels of fear in such circumstances may increase individual survival but at the expense of their long-term reproductive success. For men, the costs associated with high levels of fear may be a hinderance to their reproductive success and prevent them from capitalising on potential mating opportunities (or resources that might lead to further mating opportunities). This low fear-high risk strategy can often come at a high price however in terms of early mortality, chiefly resulting from external and avoidable causes, which are several times higher in males than females (Daly & Wilson, 1988; discussed in more depth in Chap. 2). Despite this, even if a male's risky lifestyle results in premature death, female mates (who have already invested heavily in their offspring through the mechanisms discussed earlier) can be relied upon to shoulder the burden of parenting through their continued investment in offspring, thereby securing such male's genetic contribution to the next generation.

For human women however, risk-taking has potentially negative and far-reaching effects on reproductive success. As their offspring's chief caretaker, the presence of a mother is more critical than the presence of other kin to infant survival. In a review of studies spanning 28 societies lacking access to contraception and western medical care, Sear and Mace (2008) examined the impact of kin survival on offspring survival. In every case, a mother's survival increased the likelihood of her children surviving. The percentage of children surviving a mother's death ranged from 2% to 50% with maternal effects being more marked at younger ages before children were weaned. In rural Gambia, a mother's death multiplied the odds of her child's death by 6.2 in infancy, 5.2 in toddlerhood, and 1.4 in childhood (Sear, Steele, McGregor, & Mace, 2002). Using data from seventeenth century Quebec and controlling for a range of variables including the possibility of transmitted infection and shared genetic vulnerability, Pavard, Gagnon, Desjardins, and Heyer (2005) showed that if a child's mother died in the neonatal period, the odds of the child dying were multiplied by 5.52, dropping to 1.27 if a mother dies when the child was aged 5–15 years.

Perhaps the most striking finding from Sear and Mace's (2008) review was that in 68% of cases, a father's death had no measurable impact on the survival chances of his child. It seems that the presence of a male partner has been advantageous to women, not because they increased offspring survival but because they decreased the workload on mothers. Of calories consumed in hunter-gatherer societies, between 60% (among the Nukak) and 84% (among the Ache) were contributed by men (Kaplan, Hill, Lancaster, & Hurtado, 2000). With men relieving women of the full burden of provisioning, women were better able to feed dependent children, sustain pregnancy and lactation, and return more quickly to menstrual cycling, thus shortening inter-birth intervals (Worthman, Jenkins, Stallings, & Daina, 1993) and returning them to the mate 'pool' more quickly. But the presence of a father did not dilute the central influence of the mother in keeping her children alive and, despite recent interest in human cooperative breeding, with its emphasis on the role of maternal grandmothers, their contribution to enhancing child survival is similar in magnitude to that of fathers (Mace & Sear, 2005). Mothers are the key players in ensuring their own reproductive success. Women who gestated, birthed, nursed, fed, and protected a greater number of offspring left more copies of their genes in future generations than those mothers who were less able to do so.

For a man, the minimum level of investment in offspring can be as little as a single sperm cell, thus (depending on conditions) potentially allowing them greater opportunities to sire more offspring. Conversely, there are considerable time and energy constraints on the number of offspring women can produce. Output is limited by lengthy infertile periods of gestation and lactation. It is also restricted by menopause long before the body begins to physically decline. Investment in each offspring is measured in years due to children's long and demanding period of dependency on the mother. Despite this, women's short inter-birth intervals and the potential for multiple birthing often result in the need to care simultaneously for more than one dependent child. The energetic demands on mothers and the replacement costs of each child mean that children are extremely precious. Yet childhood is the period in the human life cycle when natural selection acts most strongly (Jones, 2009). In hunter-gathering societies, an average of only 57% of children survive to age 15. The key person standing between a child and death is their chief caretaker—their mother (Hrdy, 1999).

In summary, women invest in offspring more heavily than men, early childhood is a time of heightened mortality risk for children, and mothers play the key role in ensuring their survival. Throughout history, children have been endangered by natural events and hazards (famine, drought, predation), childhood adventurousness (tall trees, cliff edges, water) and malignly motivated adult behaviour (warfare, infanticide, kidnapping). Women who had a better ability to detect and respond to danger would have left more surviving children because they were better able to monitor environmental conditions and protect their children's safety and, *by avoiding risks to themselves, they stayed alive to care for them.* If this is correct, there would have been evolutionary selection for greater fear in women, as this system would fundamentally assist in the achievement of this end.

1.3 Aims and Scope

The aim of this volume is to demonstrate how the fear system may have evolved differently in men and women because of differing selection pressures between the sexes. The underpinning argument is that a stronger fear system increases aversion to risk taking behaviour in order to keep us alive, but that this system is stronger (and more beneficial) in women than in men.

We begin by first detailing the components of the fear system and describing real world sex differences in risk taking behaviours. The chapters that follow then identify, decompose and describe those components of the fear system that may have undergone sex-linked adaptation. Each of them represents a possible target for selection where greater sensitivity in women might have conferred a net benefit in terms of reproductive success. However, the adjustment or recalibration of each of them would result in different outcomes with different potential benefits and costs. Consideration of these advantageous and disadvantageous outcomes, together with current evidence of sex differences in humans will be used to suggest where selection may have acted to enhance women's ability to keep themselves and their children alive.

References

Archer, J. (2019). The reality and evolutionary significance of human psychological sex differences. *Biological Reviews, 94*, 1381–1415.

Betzig, L. (2012). Means, variances, and ranges in reproductive success: Comparative evidence. *Evolution and Human Behavior, 33*, 309–317.

Brown, G. R., Laland, K. N., & Borgerhoff Mulder, M. (2009). Bateman's principles and human sex roles. *Trends in Ecology and Evolution, 254*, 297–304.

Campbell, A. (1999). Staying alive: Evolution, culture and intra-female aggression. *Behavioral and Brain Sciences, 22*, 203–252.

Campbell, A. (2009). What kind of selection? *Behavioral and Brain Sciences, 32*, 272–273.

Campbell, A. (2013). *A mind of her own: The evolutionary psychology of women* (2nd ed.). Oxford: Oxford University Press.
Carranza, J. (2009). Defining sexual selection as sex-dependent selection. *Animal Behaviour, 77*, 749–751.
Daly, M., & Wilson, M. (1988). *Homicide*. New York: Aldine de Gruyter.
Darwin, C. (1871/2004). *The descent of man and selection in relation to sex*. London: Penguin Classics.
Fisher, R. A. (1930). *The genetical theory of natural selection*. Oxford: Clarendon Press.
Fromhage, L., & Jennions, M. D. (2016). Coevolution of parental investment and sexually selected traits drives sex-role divergence. *Nature Communications, 7*, 12517.
Gangestad, S. W., & Simpson, J. A. (2000). The evolution of human mating: Trade-offs and strategic pluralism. *Behavioral and Brain Sciences, 23*, 573–644.
Hrdy, S. B. (1999). *Mother nature: Natural selection and the female of the species*. London: Chatto & Windus.
Jennions, M. D., & Petrie, M. (1997). Variation in mate choice and mating preferences: A review of causes and consequences. *Biological Reviews, 72*, 283–327.
Jones, J. H. (2009). The force of selection on the human life cycle. *Evolution and Human Behavior, 30*, 305–314.
Kaplan, H., Hill, K., Lancaster, J., & Hurtado, A. M. (2000). A theory of human life history evolution: Diet, intelligence, and longevity. *Evolutionary Anthropology, 9*, 156–185.
Kavanagh, E. (2006). Debating sexual selection and mating strategies. *Science, 312*, 689–697.
Kokko, H., & Jennions, M. D. (2008). Parental investment, sexual selection and sex ratios. *Journal of Evolutionary Biology, 21*, 919–948.
LeDoux, J. (1998). *The emotional brain: The mysterious underpinnings of emotional life*. New York, NY: Simon and Schuster.
Mace, R., & Sear, R. (2005). Are humans cooperative breeders? In E. Voland, A. Chasiotis, & W. Schiefenhoevel (Eds.), *Grandmotherhood: The evolutionary significance of the second half of female life* (pp. 143–159). Piscataway, NJ: Rutgers University Press.
Pavard, S., Gagnon, A., Desjardins, B., & Heyer, E. (2005). Mother's death and child survival: The case of early Quebec. *Journal of Biosocial Science, 37*, 209–227.
Rothbart, M. K., & Bates, J. E. (1998). Temperament. In W. Damon (Series Ed.) & N. Eisenberg (Vol. Ed.), *Social, emotional, and personality development: Vol. 3. Handbook of child psychology* (pp. 105–176). New York: Wiley.
Roughgarden, J., Oisha, M., & Akcay, E. (2006). Reproductive social behaviour: Cooperative games to replace sexual selection. *Science, 311*, 965–969.
Sear, R., & Mace, R. (2008). Who keeps children alive? A review of the effects of kin on child survival. *Evolution and Human Behavior, 29*, 1–18.
Sear, R., Steele, F., McGregor, A. A., & Mace, R. (2002). The effects of kin on child mortality in rural Gambia. *Demography, 39*, 43–63.
Simpson, J. A., & Gangestad, S. W. (1991). Individual differences in sociosexuality: Evidence for convergent and discriminant validity. *Journal of Personality and Social Psychology, 60*, 870–883.
Trivers, R. L. (1972). Parental investment and sexual selection. In B. Campbell (Ed.), *Sexual selection and the descent of man 1871–1971* (pp. 136–179). Chicago: Aldine.
Wilson, M., & Daly, M. (1985). Competitiveness, risk taking and violence: The young male syndrome. *Ethology and Sociobiology, 6*, 59–73.
Worthman, C. M., Jenkins, C. L., Stallings, J. F., & Daina, N. L. (1993). Attenuation of nursing-related ovarian suppression and high fertility in well-nourished, intensively breast-feeding Amele women of lowland Papua New Guinea. *Journal of Biosocial Science, 25*, 425–443.

Chapter 2
Components of the Fear System and Real-World Evidence for Sex Differences in Fear

2.1 The Fear System

We begin by briefly outlining how we detect, interpret and respond to fearful stimuli. The fear system involves numerous complex interactions between subsystems serving different functions (Adolphs, 2013; LeDoux, 2012) and it is beyond the scope of this thesis to outline every specific component (specific neuroarchitecture will only be detailed where necessary). The fear system, explained briefly, works in a series of steps. Firstly, sensory detection and identification of threat gives rise to activation of the sympathetic nervous system and automatic behavioural responses (fleeing, freezing, fighting). Continued feedback from the peripheral to the central nervous system allows for moment-to-moment fine tuning of the response for as long as the threat persists. Neurotransmitters control arousal, and connectivity within and between brain circuits. Multiple cortical regions are involved synchronously in evaluating threatening stimuli, experiencing conscious awareness of fear, the regulation of fear levels, monitoring cognitive resources and action planning. Attention is focused (and if necessary sustained) on the potential threat and memory is enhanced for contextual cues associated with it, allowing more effective responses should the threat reoccur. The subjective, phenomenological (lived) experience of fear is also an output of one or more of these processes.

Because the purpose of fear is to motivate avoidance from the potential threat, a more intense fear response would result in stronger attempts to escape from risky situations, supporting the acquisition and rehearsal of avoidant and precautionary behaviour(s). A stronger fear response would deter decisions to participate in recreational activities that might be regarded by others as exhilarating (the experience of exhilaration may in fact be a direct result of the fear system). While this conservatism has instant adaptive benefits in avoiding injury, it may also restrict exploration of behaviours that might have beneficial survival consequences beyond the immediate future (for example, willingness to seek safety by climbing or wading through water). Critical information about the potentially lethal degree of threat

A. Campbell et al., *Sex Differences in Fear Response*, SpringerBriefs in Anthropology, https://doi.org/10.1007/978-3-030-65280-7_2

conveyed by a (overly) strong fear response has the capacity to overload our normal decision making processes and may distract us from accomplishing routine cognitive and behavioural functions (short-term goal setting, planning and execution etc.).

All components of the fear system are candidates for refinement or recalibration to improve detection of and response to danger. Taking a 'reverse engineering' stance to explaining women's enhanced responsiveness to threat, there are four identifiable components in the fear system that are candidates for such adjustment (*Subjective Experience, Strength, Time Course* and *Sensitivity/Salience Detection*). We will briefly define what we mean by these components here but please note that each of these components will be discussed in more detail separately in subsequent chapters, focussing on both their likely impact on adaptive behaviour and existing evidence for a sex difference across the various experimental techniques that have been used to operationalize them.

Subjective Experience is reviewed in Chap. 3 and synthesises contemporary literature on the lived, conscious experience of fear in the presence of threatening stimuli. Here sex differences are discussed in terms of differences in levels of reported distress, phobias and self-reported measures of fear and suggests that a heightened conscious awareness of fear and threat may either be an adaptive mechanism or a consequence of the components of an adaptive fear system.

Chapter 4 discusses sex differences in the relative *Strength* of the fear response system, which while overlapping with conscious experience, focuses more specifically on the relative activity of brain areas implicated in the fear system. Here we discuss evidence of increased activity for females in areas such as the Amygdala, the Orbitofrontal Cortex and the Anterior Cingulate Cortex which may increase deterrence and desistance from dangerous, risky activities and enhance survival.

Chapter 5 proposes that women may experience fear over a protracted *Time Course* compared to men, as well as spend more time processing and remaining vigilant around potentially threatening stimuli. Evidence from studies of both classical and operant conditioning in relation to fear extinction are discussed alongside other sources of empirical evidence. Here we suggest that extended processing times and heightened vigilance around threatening stimuli would play a role in keeping women alive.

The final component, discussed in Chap. 6, pertains to sex differences in the *Sensitivity* to (i.e. awareness of the presence of a potential threat) and S*alience* of (i.e. the evaluation of the level of danger the potential threat poses) fear provoking stimuli. Evidence for sex differences in relation to each concept is discussed separately alongside the relative costs and benefits of having increased sensitivity and salience within an integrated fear system.

In Chap. 7, we summarise the components of the fear system and their interconnectivity in relation to the evidence discussed in the preceding chapters as well as suggest fruitful avenues of approach for future research considering various limitations of the current literary body.

Appendix A presents a summary of the experimental techniques (of which there are many) reviewed in relation to each of the aforementioned components. This is included here as a quick reference guide to the types of tasks usually employed to explore the psychology of fear. Similarly, Appendix B presents a short-hand summary of the role of various neuroanatomic regions believed critical in the fear system. We recommend readers refer to these tables when considering the empirical studies discussed within each chapter.

2.2 Risk Taking and Real World Sex Differences

Before we examine each component however, it is useful to begin by discussing sex differences in psychological domains closely aligned with the fear system readily observable in both real world and laboratory settings. Because fear is, among other things (Adolphs, 2013), a subjective emotion unavailable for objective measurement, behavioural risk taking constitutes a useful proxy measure. We made the claim earlier that higher levels of fear would lead to increased aversion towards risky activities. Risk avoidance is indeed associated with fear and increased fear would be expected to deter women from participation in risky behaviours given the potentially injurious or even fatal consequences to their overall fitness (i.e. their own future reproduction or the future fitness of their offspring). There is ample evidence to support the link between fear and risk aversion. Habib, Cassotti, Moutier, Houde, and Borst (2015) demonstrate using a computerised gambling task that invoking fearful stimuli (fearful faces) significantly increased risk aversion. Variations of this paradigm also show this effect is replicable in non-western populations (She, Eimontaite, Zhang, & Sun, 2017). Fear increases perceptions of uncertainty and a lowered sense of self control, two factors that greatly increase the likelihood of risk aversive strategies (Lerner & Keltner, 2000). A recent meta-analysis (Wake, Wormwood, & Satpute, 2020) across 136 effect sizes confirms a small to moderate effect of fear on increasing risk estimation and decreasing risk taking ($d = 0.22$, see Appendix C for guidance on the interpretation of effect sizes). The effect of fear on risk taking is stronger when the rewards are real compared to hypothesised ($d = 0.30$). While there is variability in the effect size that varies by method, manipulation, paradigm etc., comparatively few studies in the analysis found the reverse to be true. There thus appears to be a consensus in the literature that fear increases desistance from risky activities.

If risk is therefore an adequate and measurable proxy for fear, is there evidence that the sexes differ across different risk-taking scenarios? Perhaps the most striking evidence for such a sex difference comes from officially recorded statistics on death and injury resulting from risky activities. Men die from external causes more frequently than women, with a peak of 4:1 in the male-female ratio between the ages of 20 and 24 (Kruger & Nesse, 2006). In this age group, men are over six times more likely than women to commit suicide, five times more likely to die from non-vehicle accidents (including falls, drowning, choking, electrocution, firearm accidents, and fires: Pampel, 2001), four times more likely to be the victim of homicide, and three times more likely to die in traffic accidents. Over 97% of dangerous driving offences are committed by men (Beattie, 2008; Norris, Matthews,

& Riad, 2000). Men make up 74% of those attending hospital Accident and Emergency Departments because of violence-related incidents (Sivarajasingam et al., 2014). Men constitute 76% of all criminal arrests in the United States, committing 89% of homicides and 82% of all violent crime (Federal Bureau of Investigation, 2012). Worldwide, men use drugs (alcohol, tobacco, cannabis, and cocaine) more than women (Degenhardt et al., 2008). They participate more often in extreme sports, such as skydiving and mountain climbing (Robinson, 2008). Even in the less risky outdoor pursuit of swimming, they are more likely than women to drown (Howland, Hingson, Mangione, & Bell, 1996).

Risky behaviour can also be assessed in the lab. A lab measure of risk-taking which correlates with real-world risky behaviours such as smoking is the Balloon Analogue Risk Task (BART), which assesses a participant's willingness to risk monetary loss in the pursuit of winning a bigger reward. A meta-analysis of 10 studies using the BART (Cross, Copping, & Campbell, 2011) found that men scored consistently higher than women ($d = 0.36$). In a meta-analysis of sex differences in risk taking, Byrnes, Miller, and Schafer (1999) found greater risk taking by men over a range of tasks ($d = 0.19$), and these were most marked in studies involving real rather than hypothetical risk. They noted that "the processes involved in the translation of cognitions to behaviors (e.g., fear responses) may explain gender differences in risk taking more adequately than the cognitive processes involved in the reflective evaluation of options" (Byrnes et al., 1999, p. 378). In other words, the sex difference is driven by an emotional response to risk (i.e. fear) rather than simply the awareness that something is risky in its own right.

Aggression is a high-risk behaviour and laboratory estimates of sex differences in aggression are considerably larger than those found by Byrnes et al. (1999). Under the same experimental treatment, men exceed women in physical aggression (effect sizes between $d = 0.59$ and $d = 0.91$) and verbal aggression (between $d = 0.28$ and $d = 0.46$) (Knight, Fabes, & Higgins, 1996; Knight, Guthrie, Page, & Fabes, 2002). This sex difference is mediated by women's greater fear (Bettencourt & Miller, 1996; Eagly & Steffen, 1986; Hyde, 1986). Naturalistic observational studies confirm greater male aggression at all ages and across all cultures sampled (Archer, 2004), as do violent crime statistics (see above).

Risk tolerance has also been measured using driving simulators. Young men drive more riskily in terms of speeding and crossing the central white line than women do (Taubman-Ben-Ari, Eherenfreund-Hager, & Prato, 2016). When making left-hand turns against oncoming traffic, male drivers accept smaller gaps than female drivers (Yan, Radwan, & Guo, 2007). Men overtake more often on two-lane roads, drive closer to the vehicle in front, tolerate a shorter delay before beginning the overtaking manoeuvre, have a lower critical overtaking gap than females' and complete the overtaking manoeuvre more quickly (Farah, 2011). As pedestrians, men more often cross roads when it is unsafe to do so (Pawlowski, Rajinder, & Dunbar, 2008) and as cyclists, fail to use lights and stop at railroad crossings (Cobey, Stulp, Laan, Buunk, & Pollet, 2013).

The behavioural approach task is a lab-based observational tool that represents fear as an ordinal (rank) score measuring willingness to approach and contact a feared stimulus. Because aversion to spiders is common, they have been used to assess fear in undergraduate samples. Participants' approach behaviour is scored on a stepwise Guttman-type scale (where each successive step evokes more fear than the previous) from 1 (entering the room where a tarantula is) to 12 (allowing it to crawl on their hand or touching it for 3 seconds with 2 fingers). Women (McLean & Hope, 2010) and girls aged 9–12 years (Klein, Becker, & Rinck, 2011) take significantly fewer approach steps than boys.

In summary, not only can we infer that fear and risk are linked, but publicly available statistics on mortality, morbidity, offending, and public health robustly confirm men's greater risk taking. These differences are supported by systematic observations in laboratory settings that attest to men's greater willingness to use direct aggression, to drive recklessly and to confront feared stimuli. The fact that women consistently engage less in risky behaviours can be taken as evidence that an underlying sex difference in fear is potentially responsible for encouraging survival via avoidance of threatening stimuli. While this body of work is compelling, it alone does not describe the mechanism or mechanisms underpinning sex differences in the fear system. Attention in the forthcoming chapters turns to a more detailed analysis of sex differences emerging in the fear system components identified earlier in this chapter.

References

Adolphs, R. (2013). The biology of fear. *Current Biology, 23*, R79–R93.

Archer, J. (2004). Sex differences in aggression in real-world settings: A meta-analytic review. *Review of General Psychology, 8*(4), 291–322.

Beattie, G. (2008). *Sex differences in driving and insurance risk: Understanding the neurobiological and evolutionary foundations of the differences*. Manchester: University of Manchester. Retrieved from http://www.esure.com/wcm/groups/public/documents/webcontent/wcmprod_sex_differences.pdf.

Bettencourt, B. A., & Miller, N. (1996). Gender differences in aggression as a function of provocation: A meta-analysis. *Psychological Bulletin, 119*, 422–447.

Byrnes, J. P., Miller, D. C., & Schafer, W. D. (1999). Gender differences in risk taking: A meta-analysis. *Psychological Bulletin, 125*, 367–383.

Cobey, K. D., Stulp, G., Laan, F., Buunk, A. P., & Pollet, T. V. (2013). Sex differences in risk taking behavior among Dutch cyclists. *Evolutionary Psychology, 11*(2), 350–364.

Cross, C. P., Copping, L. T., & Campbell, A. (2011). Sex differences in impulsivity: A meta-analysis. *Psychological Bulletin, 137*(1), 97–130.

Degenhardt, L., Chiu, W.-T., Sampson, N., Kessler, R. C., Anthony, J. C., Angermeyer, M., et al. (2008). Toward a global view of alcohol, tobacco, cannabis, and cocaine use: Findings from the WHO World Mental Health Surveys. *PLoS Medicine, 5*(7), e141.

Eagly, A. H., & Steffen, V. J. (1986). Gender and aggressive behavior: A meta-analytic review of the social psychological literature. *Psychological Bulletin, 100*, 309–330.

Farah, H. (2011). Age and gender differences in overtaking manoeuvre on two-lane rural highways. *Transportation Research, 2248*, 30–36.

Federal Bureau of Investigation. (2012). *Crime in the United States 2011*. https://www.fbi.gov/about-us/cjis/ucr/crime-in-the-u.s/2011/crime-in-the-u.s.-2011/tables/table-33

Habib, M., Cassotti, M., Moutier, S., Houde, O., & Borst, G. (2015). Fear and anger have opposite effects on risk seeking in the gain frame. *Frontiers in Psychology, 6*, 253.

Howland, J., Hingson, R., Mangione, T. W., & Bell, N. (1996). Why are most drowning victims men? Sex differences in aquatic skills behaviors. *American Journal of Public Health, 86*, 93–96.

Hyde, J. S. (1986). Gender differences in aggression. In J. S. Hyde & M. C. Linn (Eds.), *The psychology of gender: Advances through meta-analysis* (pp. 51–66). Baltimore, MD: Johns Hopkins University Press.

Klein, A. K., Becker, E. S., & Rinck, M. (2011). Approach and avoidance tendencies in spider fearful children: The Approach-Avoidance Task. *Journal of Child and Family Studies, 20*, 224–231.

Knight, G. P., Fabes, R. A., & Higgins, D. A. (1996). Concerns about drawing causal inferences from meta-analyses: An example in the study of gender differences in aggression. *Psychological Bulletin, 119*, 410–421.

Knight, G. P., Guthrie, I. K., Page, M. C., & Fabes, R. A. (2002). Emotional arousal and gender differences in aggression: A meta analysis. *Aggressive Behavior, 28*, 366–393.

Kruger, D. J., & Nesse, R. M. (2006). An evolutionary life-history framework for understanding sex differences in human mortality rates. *Human Nature, 17*, 74–97.

LeDoux, J. (2012). Rethinking the emotional brain. *Neuron, 73*, 653–676.

Lerner, J. S., & Keltner, D. (2000). Beyond valence: Toward a model of emotion-specific influences on judgment and choice. *Cognition and Emotion, 14*, 473–493.

McLean, C. P., & Hope, D. A. (2010). Subjective anxiety and behavioral avoidance: Gender, gender role, and perceived confirmability of self-report. *Journal of Anxiety Disorders, 24*, 494–502.

Norris, F. H., Matthews, B. A., & Riad, J. K. (2000). Characterological, situational, and behavioral risk factors for motor vehicle accidents: A prospective examination. *Accident Analysis and Prevention, 32*(4), 505–515.

Pampel, F. C. (2001). Gender equality and the sex differential in mortality from accidents in high income nations. *Population Research and Policy Review, 20*, 397–421.

Pawlowski, B., Rajinder, A., & Dunbar, R. I. M. (2008). Sex differences in everyday risk-taking behavior in humans. *Evolutionary Psychology, 6*(1), 29–42.

Robinson, V. (2008). *Everyday masculinities and extreme sports: Male identity and rock climbing*. Oxford: Berg.

She, S., Eimontaite, I., Zhang, D., & Sun, Y. (2017). Fear, anger and risk preference reversals: An experimental study on a Chinese sample. *Frontiers in Psychology, 8*, 1371.

Sivarajasingam, V., Page, N., Morgan, P., Matthews, K., Moore, S., & Shepherd, J. (2014). Trends in community violence in England and Wales 2005-2009. *Injury, 45*(3), 592–598.

Taubman-Ben-Ari, O., Eherenfreund-Hager, A., & Prato, C. G. (2016). The value of self-report measures as indicators of driving behaviors among young drivers. *Transportation Research Part F: Traffic Psychology and Behaviour, 39*, 33–42.

Wake, S., Wormwood, J., & Satpute, A. B. (2020). The influence of fear on risk taking: A meta-analysis. *Cognition and Emotion, 34*(6), 1143–1159. https://doi.org/10.1080/02699931.2020.1731428.

Yan, X., Radwan, E., & Guo, D. (2007). Effects of major-road vehicle speed and driver age and gender on left-turn gap acceptance. *Accident Analysis and Prevention, 39*(4), 843–852.

Chapter 3
Sex Differences in the Subjective Experience of Fear

While the evolutionary significance and source of consciousness remains controversial, fear is a state that is experienced consciously, and women's experience of fear might be more accessible to awareness than men's. This might be significant only in the extent to which it reflects a stronger response in the underlying fear system (see Chap. 4, Strength, below) or it might be significant because the conscious experience of fear feeds into decision-making (Loewenstein, Weber, Hsee, & Welch, 2001). Many studies have examined sex differences in the experience of fear using self-report methods: such studies might simply ask a large number of men and women to report how frequently and how intensely they experience fear. National and international surveys of this type show that women report more intense and long-lasting fear than men (Brebner, 2003; Brody & Hall, 1993; Crawford, Kippax, Onyx, Gault, & Benton, 1992; Fischer, 1993; Fischer & Manstead, 2000; Fischer, Mosquera, van Vianen, & Manstead, 2004; Gullone, 2000; McLean & Anderson, 2009). Simple surveys such as this allow data to be gathered from large samples, but the use of single-item measures can often lead to questions around validity (i.e. how well they actually reflect the real subjective experience of fear). Questionnaire measures are most effective when they are composed of multiple items that have been previously empirically validated (see Rust & Golombok, 2014 for how this can be achieved empirically).

Sex differences in susceptibility to phobias (a marked and persistent fear of specific objects or situations, which the sufferer recognises to be excessive or unreasonable) are instructive. The most common phobias are thought to correspond to specific dangers faced by humans during their evolutionary past, such as dangerous animals, heights, or enclosed spaces (and not, for example, guns or busy roads). Nesse (1997, p. 64) suggests that phobias might exist because "repeated false alarms may cost less than a single failure to respond when the danger is great." Women are more likely than men to have phobias, but only of things which threaten bodily integrity; blood, animals, or medical procedures. Women do not have higher levels of social phobia than men (Marks, 1987). A widely used multi-item measure of fear is the Fear Survey Schedule (FSS) which assesses phobias. The FSS-II, developed

A. Campbell et al., *Sex Differences in Fear Response*, SpringerBriefs in Anthropology, https://doi.org/10.1007/978-3-030-65280-7_3

for research purposes routinely demonstrates higher scores for women, with effect sizes of between $d = -0.70$ (Geer, 1965) and $d = -0.76$ (Bernstein & Allen, 1969). A more recent inventory of specific phobias (McCraw & Valentiner, 2015) decomposed fear responses into five categories (desire to escape or avoid; feelings of fear or anxiety; physiological symptoms; fear of losing control; risk analysis). Women scored significantly higher than men in each of these categories and in overall score ($d = -0.59$). Women score higher than men on the Fear of Pain Questionnaire (FPQ-III; McNeil & Rainwater, 1998), with effect sizes of $d = -0.65$ for students, $d = -0.59$ for medical patients, and $\mathrm{d} = -0.86$ for chronic pain patients. The sex difference in scores on the Fear of Crime (FOC) Scale, $\mathrm{d} = -0.39$, makes gender the strongest known predictor of FOC scores—stronger even than previous experience of victimisation (Collins, 2016). Large effect sizes have also been found for the Harm Avoidance scale of the Multidimensional Personality Questionnaire, which assesses the preference for a safe but tedious activity over a stimulating but risky one ($d = -0.78$; Cross, Copping, & Campbell, 2011), and the Fearlessness scale of Psychopathic Personality Inventory, which measures an eagerness for risk and an absence of fear ($d = 0.76$; Lee & Salekin, 2010).

A key question is whether sex differences in self-report measures are methodological artefacts of gendered self-presentation resulting from men's reluctance to admit fear. There is some evidence that gendered expectations influence responses to fear questionnaires: Fear scores are negatively associated with masculinity (Arrindell, 2000) and positively with femininity (Tucker & Bond, 1997). Nevertheless, studies which control for gender role still find a significant effect of biological sex (Arrindell, Kolk, Pickersgill, & Hageman, 1993; Arrindell, Mulkens, Kok, & Vollenbroek, 1999; Campbell et al., 2016; Dillon, Wolf, & Katz, 1985). Furthermore, the sex difference in reported fear is not eliminated by the 'bogus pipeline' technique, in which respondents are (falsely) instructed that the researchers are monitoring physiological indictors that will allow them to spot dishonest responses (McLean & Hope, 2010; Pickersgill & Arrindell, 1994; Stoyanova & Hope, 2012). Questionnaire measures can also be developed using Rasch scaling to ensure that items which are biased towards one sex are removed from the final measure. One such inventory, developed recently by Campbell et al. (2016), showed greater fear in women with effect sizes ranging between $d = -1.04$ (UK) and $d = -1.23$ (Romania).

Sex differences in self-reported fear are often confounded with sex differences in everyday experiences. A solution to this is to conduct laboratory studies in which participants are asked to rate their fear in response to the same stimulus. A higher rating by women suggests either a stronger conscious experience of fear or a greater willingness to report it. Two studies which tested participants breathing high-CO_2 air (see Appendix A) found no significant sex differences in rated anxiety, fear, or panic symptoms (Griez, Colasanti, van Diest, Salamon, & Schruers, 2007; Leibold et al., 2013), as did three studies which employed only the highest dosage (35% CO_2) (Nardi et al., 2007; Perna et al., 2004; Poonai et al., 2000). However, in some studies women rated their fear higher than men and/or experienced the event as more aversive (Bunaciu, Feldner, Babson, Zvolensky, & Eifert, 2012; Forsyth & Eifert, 1998; Kelly & Forsyth, 2007; Kelly, Forsyth, & Karekla, 2006; Nillni, Berenz, Rohan, & Zvolensky, 2012). On the behavioural approach task (discussed earlier), Stoyanova and Hope (2012) found women's self-rated distress was higher than men's after approaching the feared stimulus (in this case, a spider) but not before. Furthermore, McLean and Hope (2010) found that women scored higher both prior to and immediately after the task suggesting greater experience of anticipatory as well as reactive fear. Taken together, these experiments tentatively suggest that women have a stronger subjective experience of fear than men.

Researchers have also examined sex differences in fear following exposure to man-made and natural disasters. Between 3 and 6 months after the New York 9/11 terrorist attacks, a stratified random sample of residents was asked about their current mental health status. Women scored significantly higher than men on a trauma scale and all three of its subscales: intrusion, avoidance/numbing and hyperarousal (DeLisi et al., 2003). A meta-analysis conducted by Tolin and Foa (2006) showed that across different categories of traumatic events (natural disasters, nonsexual assault, illness and injury etc.) women and girls were more likely to show greater frequency or severity of PTSD symptoms than men and boys ($d = -0.29$). They were also more likely to meet the criteria for PTSD despite being less likely to experience such traumatic events. Even in more naturalistic settings, it would thus appear that women self-report higher levels of fear than men (although issues of self-report remain).

In summary, women consistently report experiencing higher levels of fear on self-report psychometric scales. Although masculine gender role is negatively correlated with self-reported fear, controlling for gender role statistically or by experimental manipulation, does not eliminate the sex difference in reported fear. When exposed to the same fear-inducing situations, whether public disasters or behavioural tasks, women tend to report greater distress. However, when physiological panic is induced in the laboratory sex differences are less evident, perhaps because of a ceiling effect (i.e. the fear response is extreme for everyone, irrespective of sex). Whether this heightened awareness of fear is advantageous to reproductive fitness in its own right, or simply a by-product of the remaining components of an adaptive fear system (forthcoming) also remains to be seen.

References

Arrindell, W. A. (2000). Phobic dimensions: IV. The structure of animal fears. *Behaviour Research and Therapy, 38*(5), 509–530.

Arrindell, W. A., Kolk, A. M., Pickersgill, M. J., & Hageman, W. J. (1993). Biological sex, sex role orientation, masculine sex role stress, dissimulation and self-reported fears. *Advances in Behaviour Research and Therapy, 15*, 103–146.

Arrindell, W. A., Mulkens, S., Kok, J., & Vollenbroek, J. (1999). Disgust sensitivity and the sex difference in fears to common indigenous animals. *Behaviour Research and Therapy, 37*(3), 273–280.

Bernstein, D. A., & Allen, G. J. (1969). Fear survey schedule (II): Normative data and factor analyses based on a large college sample. *Behaviour Research and Therapy, 7*, 403–407.

Brebner, J. (2003). Gender and emotions. *Personality and Individual Differences, 34*, 387–394.

Brody, L. R., & Hall, J. A. (1993). Gender and emotion. In M. Lewis & J. M. Haviland (Eds.), *Handbook of emotions* (pp. 447–460). New York, NY: Guildford Press.

Bunaciu, L., Feldner, M. T., Babson, K. A., Zvolensky, M. J., & Eifert, G. H. (2012). Biological sex and panic-relevant anxious reactivity to abrupt increases in bodily arousal as a function of biological challenge intensity. *Journal of Behavior Therapy and Experimental Psychiatry, 43*, 526–531.

Campbell, A., Coombes, C., David, R., Opre, A., Grayson, L., & Muncer, S. (2016). Sex differences are not attenuated by a sex-invariant measure of fear: The Situated Fear Questionnaire. *Personality and Individual Differences, 97*, 210–219.

Collins, R. E. (2016). Addressing the inconsistencies in fear of crime research: A meta-analytic review. *Journal of Criminal Justice, 47*, 21–31.

Crawford, J., Kippax, S., Onyx, J., Gault, U., & Benton, P. (1992). *Emotion and gender: Constructing meaning from memory*. London: Sage.

Cross, C. P., Copping, L. T., & Campbell, A. (2011). Sex differences in impulsivity: A meta-analysis. *Psychological Bulletin, 137*(1), 97–130.

DeLisi, L. E., Maurizio, A., Yost, M., Papparozzi, C. F., Fulchino, C., Katz, C. L., . . . Stevens, P. (2003). A survey of New Yorkers after the Sept. 11, 2001, terrorist attacks. *American Journal of Psychiatry, 160*, 780–783.

Dillon, K. M., Wolf, E., & Katz, H. (1985). Sex roles, gender, and fear. *Journal of Psychology, 119*, 355–359.

Fischer, A. H. (1993). Sex differences in emotionality: Fact or stereotype? *Feminism and Psychology, 3*, 303–318.

Fischer, A. H., & Manstead, A. S. R. (2000). Gender and emotions in different cultures. In A. H. Fischer (Ed.), *Gender and emotion: Social psychological perspectives* (pp. 71–94). Cambridge: Cambridge University Press.

Fischer, A. H., Mosquera, P. M. R., van Vianen, A., & Manstead, A. S. R. (2004). Gender and culture differences in emotion. *Emotion, 4*, 87–94.

Forsyth, J. P., & Eifert, G. H. (1998). Response intensity in content-specific conditioning comparing 20% vs. 13% CO2-enriched air as unconditioned stimuli. *Journal of Abnormal Psychology, 107*, 291–304.

Geer, J. H. (1965). The development of a scale to measure fear. *Behaviour Research and Therapy, 3*, 45–53.

Griez, E. J., Colasanti, A., van Diest, R., Salamon, E., & Schruers, K. (2007). Carbon dioxide inhalation induces dose-dependent and age-related negative affectivity. *PLoS One, 2*, e987.

Gullone, E. (2000). The development of normal fear: A century of research. *Clinical Psychology Review, 20*, 429–451.

Kelly, M. M., & Forsyth, J. P. (2007). Sex differences in response to an observational fear conditioning procedure. *Behavior Therapy, 38*, 340–349.

Kelly, M. M., Forsyth, J. P., & Karekla, M. (2006). Sex differences in response to a panicogenic challenge procedure: An experimental evaluation of panic vulnerability in a non-clinical sample. *Behaviour Research and Therapy, 44*, 1421–1430.

Lee, Z., & Salekin, R. T. (2010). Psychopathy in a noninstitutional sample: Differences in primary and secondary subtypes. *Personality Disorders: Theory, Research, and Treatment, 1*(3), 153–169.

Leibold, N. K., Viechtbauer, W., Goossens, L., De Cort, K., Griez, E. J., Myin-Germeys, I., . . . Schruers, K. R. J. (2013). Carbon dioxide inhalation as a human experimental model of panic: The relationship between emotions and cardiovascular physiology. *Biological Psychology, 94*, 331–340.

Loewenstein, G. F., Weber, E. U., Hsee, C. K., & Welch, N. (2001). Risk as feelings. *Psychological Bulletin, 127*(2), 267–286.

Marks, I. M. (1987). Fears, phobias, and rituals: Panic, anxiety, and their disorders. Oxford University Press on Demand.

McCraw, K. S., & Valentiner, D. P. (2015). The circumscribed fear measure: Development and initial validation of a trans-stimulus phobia measure. *Psychological Assessment, 27*(2), 403–414.

McLean, C. P., & Anderson, E. R. (2009). Brave men and timid women? A review of the gender differences in fear and anxiety. *Clinical Psychology Review, 29*, 496–505.

McLean, C. P., & Hope, D. A. (2010). Subjective anxiety and behavioral avoidance: Gender, gender role, and perceived confirmability of self-report. *Journal of Anxiety Disorders, 24*, 494–502.

McNeil, D. W., & Rainwater, A. J. (1998). Development of the fear of pain questionnaire-III. *Journal of Behavioral Medicine, 21*(4), 389–410.

Nardi, A. E., Valenca, A. M., Lopes, F. L., de Melo-Neto, V. L., Freire, R. C., Veras, A. B., . . . Zin, W. A. (2007). Caffeine and 35% carbon dioxide challenge tests in panic disorder. *Human Psychopharmacology, 22*, 231–240.

Nesse, R. M. (1997). An evolutionary perspective on panic disorder and agoraphobia. In S. Baron-Cohen (Ed.), *The maladapted mind: Classic readings in evolutionary psychopathology* (pp. 73–85). Hove: Psychology Press.

Nillni, Y. I., Berenz, E. C., Rohan, K. J., & Zvolensky, M. J. (2012). Sex differences in panic-relevant responding to a 10% carbon dioxide-enriched air biological challenge. *Journal of Anxiety Disorders, 26*, 165–172.

Perna, G., di Bella, D., Favaron, E., Cucchi, M., Liperi, L., & Bellodi, L. (2004). Lack of relationship between CO2reactivity and serotonin transporter gene regulatory region polymorphism in panic disorder. *American Journal of Medical Genetics Part B: Neuropsychiatric Genetics, 129B*, 41–43.

Pickersgill, M. J., & Arrindell, W. A. (1994). Men are innocent until proven guilty: A comment on the examination of sex differences by Pierce and Kirkpatrick (1992). *Behavior Research and Therapy, 32*(1), 21–28.

Poonai, N., Antony, M. M., Binkley, K. E., Stenn, P., Swinson, R. P., Corey, P., . . . Tarlo, S. (2000). Carbon dioxide inhalation challenges in idiopathic environmental intolerance. *Journal of Allergy and Clinical Immunology, 105*, 358–363.

Rust, J., & Golombok, S. (2014). *Modern psychometrics: The science of psychological assessment*. London: Routledge.

Stoyanova, M., & Hope, D. A. (2012). Gender, gender roles, and anxiety: Perceived confirmability of self report, behavioral avoidance, and physiological reactivity. *Journal of Anxiety Disorders, 26*, 206–214.

Tolin, D. F., & Foa, E. B. (2006). Sex differences in trauma and posttraumatic stress disorder: A quantitative review of 25 years of research. *Psychological Bulletin, 132*(6), 959–992.

Tucker, M., & Bond, N. W. (1997). The roles of gender, sex role, and disgust in fear of animals. *Personality and Individual Differences, 22*, 135–138.

Chapter 4
Sex Differences in Strength of Fear Response

While the strength of the fear system can be inferred in part from the conscious experience of fear, it is also instructive to measure the activity of the relevant brain areas directly. Enhanced excitatory activation of some or all of these areas might be expected to increase the intensity of fear. The strength of the fear system has been assessed chiefly via neuroimaging studies in which images of faces showing different emotions (including fear and anger) are presented and regional activation in the viewer's brain is recorded. Most studies identify a priori regions of interest, but increasingly sophisticated statistical programs and imaging techniques allow researchers to investigate structural, functional and temporal connectivity within and between brain circuits. The interpretation of findings is complicated by the difficulty (some would say impossibility) of mapping the correspondence between structure and function. For example, signals reflecting changes in blood-oxygen levels (so called BOLD signals) indicate regions that are active in response to threat, but these regions could be registering threat, augmenting the fear response, suppressing the fear response, supporting decision-making processes, or a number of other concurrent functions. Nonetheless, some key circuits associated with fear have consistently been identified. These include sub-cortical structures such as the amygdala (located within the medial temporal lobe and associated with fear registration) and the hypothalamus (located at the base of the brain, guiding our autonomic responses). Areas of the mid-brain such as the periaqueductal grey have a role in the expression of behavioural fear while broader regions such as the ventromedial and orbitofrontal cortices show heightened activity in relation to fear regulation, task monitoring, future simulation and action decisions. Appendix B provides more specific details regarding the various neural structures covered within the chapters of this text and their associations with the components of the fear system.

Only a few neuroimaging studies of fear have directly compared males and females in the same experiment. Several meta-analyses however, have examined sex differences in the activation of specific brain structures in response to 'emotional' faces by, for example, comparing the size of effects from studies with differing proportions of men and women. Wager, Phan, Liberzon, and Taylor

A. Campbell et al., *Sex Differences in Fear Response*, SpringerBriefs in Anthropology, https://doi.org/10.1007/978-3-030-65280-7_4

(2003) found greater activation across multiple areas of the brain in women, including the subcallosal anterior cingulate, the thalamus, midbrain and cerebellum. These patterns of activation were attributed to women's greater attention to their sensory and emotional state. Stevens and Hamann's (2012) meta-analysis grouped studies by stimulus valence. For negative stimuli, women showed greater left amygdala and hippocampus activation, suggestive of enhanced memory encoding, and greater anterior cingulate and medial prefrontal cortex activity reflecting a focus on internal mental state or attempts to modulate it. Sergerie, Chochol, and Armony (2008) however found no sex difference in the extent of amygdala activity relative to all brain activations.

Unfortunately, none of these analyses addressed sex differences in specific emotions such as fear. Categories such as 'negative' or 'aversive' stimuli conflate regional activations to emotions as disparate as anger, disgust, guilt, sadness and fear. Although meta-analyses are useful in addressing problems of small sample size and low power in neuroimaging studies, they are limited in their ability to identify sex differences in specific structures in response to specific types of stimuli. In addition, because many studies use participants of only one sex, meta-analyses compare men and women over different experimental designs, task instructions and emotional stimuli. We must therefore be cautious in our interpretation of such analyses when extending to fear. Despite this limitation however, there is evidence from other experimental studies (forthcoming) that corroborates the importance of many of these brain structures in relation to threatening, fear evoking stimuli.

Before turning to the few studies which have specifically focused on sex differences in response to threat, there is a methodological issue that complicates interpretation in emotional processing studies generally. Participants view a series of faces expressing different emotions and are instructed to respond to some task-irrelevant aspect of the image (e.g. sex). Researchers rarely make explicit the psychological processes that are assumed to be taking place when the viewer is presumably decoding the expressed emotion. This may involve an emotional response that duplicates the expressed emotion (Wood, Rychlowska, Korb, & Niedenthal, 2016) but it is equally possible that it differs from the emotion being expressed (Davis et al., 2011). Emotional processing of a fearful face may involve one or all of the following: the identification of facial features associated with a person experiencing fear; a subjective ('mirroring') experience of fear; or a 'reactive' emotion (such as dominance, concern or curiosity) elicited by the sight of a frightened person. Regarding this last 'reactive' possibility, it has now been established that a fearful facial expression signals vulnerability and need for assistance (Marsh & Ambady, 2007; Marsh, Ambady, & Kleck, 2005). Although it has been widely assumed that a fearful face communicates threat in the locality of the poser and therefore prompts an avoidant response in the viewer, fearful faces may actually elicit approach responses instead (Hammer & Marsh, 2015). Hence an angry face, rather than a fearful one, may be more effective in evoking subjective fear in the viewer. This may be especially true for women (see below).

Several neuroimaging studies suggests a sex difference in the response and decay rate of the amygdala in the presence of threatening stimuli. Thomas et al. (2001)

found that younger boys showed decreases in amygdala activation while activity in girls remained stable with repeated presentations of fearful faces. This result has been replicated in adults by Williams et al. (2005), with men showing a reduction of right amygdala activation between the early and late stages of the experiment while women's right amygdala activation increased both in terms of extent (voxel number) and effect size (standardized magnitude of difference between control and fear conditions). Skin conductance (electrical activity across the skin which is indicative of arousal caused by the sympathetic nervous system) followed a similar pattern, decreasing in men but not women. Taken together, these results imply that in the presence of fear evoking stimuli, the strength of amygdala activation decays in males while remaining at least constant in females. Furthermore, familiarity with stimuli, does not dampen the amygdala activation of women as strongly as it does that of men. In a study that explicitly manipulated stimulus novelty versus familiarity, as well as using positive and negative emotional stimuli (Andreano, Dickerson, & Barrett, 2014), women demonstrated a pattern of greater amygdala response than men specifically to familiar, negative, stimuli. Women showed smaller differences between novel and familiar images compared to men in several other brain regions including the Anterior Insula, Orbitofrontal Cortex, the hippocampus and pregenual Anterior Cingulate Cortex. Finally, this perseverative activation was positively correlated with self-rated anxiety. The culmination of these results suggests that sex differences in the strength of the fear response are related to differences in the activity (and attenuation thereof) of the amygdala.

McClure et al. (2004) found that women showed greater neural specificity in their response to threatening stimuli than men, although this effect was not found in early adolescents. Neutral, happy, fearful and angry emotional facial expressions were used as stimuli, with a particular focus on the comparison between fearful faces (referred to as 'ambiguous threat' because they were assumed to signal the indirect presence of threat in the locality of the actor) and angry faces (referred to as 'unambiguous threat' since the gaze was directed unambiguously at the viewer). Women showed a pattern of enhanced activity in areas previously indicated to be part of the fear circuitry, specifically in those contrasts involving the angry face. Women showed greater activity than men in the right amygdala and right OFC to the angry versus neutral face. As suggested above, the presentation of an angry face appears to be more fear-provoking than a fearful face, at least to adult women. Furthermore, in the contrast between angry and fearful faces, they showed greater bilateral amygdala and right Orbitofrontal Cortex activity to the angry faces. Compared to baseline fixation, men showed a more diffuse pattern of activation with increased excitation in at least one region to all images except the happy face. By contrast, women's reactivity was confined to the angry face condition which activated the Orbitofrontal Cortex, amygdala and Anterior Cingulate Cortex. The sexes were therefore also distinguishable by the selectivity of fear circuit activation in the presence of threat.

Several researchers have proposed that studies of negative emotion processing support a distinction between women's focus on internal feeling states and men's focus on external sensory data and their implications for action (Wager et al., 2003).

This was supported by Lungu, Potvin, Tikasz, and Mendrek (2015) who established that functional connectivity between the right amygdala and the dorsomedial prefrontal cortex in response to highly arousing negative images was not only stronger in men, but was positively correlated with testosterone and negatively correlated with subjective ratings of emotional intensity and self-rated femininity. Previous research suggests that the dorsomedial prefrontal cortex has a role in threat appraisal and response selection (Etkin, Egner, & Kalisch, 2011) and this may suggest that men have a stronger action-orientation to emotional processing while women attend more strongly to the feelings elicited by the stimuli. Moriguchi, Touroutoglou, Dickerson, and Barrett (2014) identified a similar distinction in terms of attentional focus. Although men and women did not differ in the intensity of self-rated moment-to-moment subjective arousal to emotional pictures, women's arousal was accompanied by increased activity in the Anterior Insula implicated in interoception (awareness of one's own internal states) and the subjective experience of emotion. Men's arousal was correlated with activity in the primary visual cortex and fusiform gyrus (part of the ventral visual stream) suggesting an exterior focus during emotional arousal. They suggest that for women, emotion is experienced as an internal somatic state while for men emotion is 'externalised' into the visual environment. The functional connection between the Anterior Insular Cortex and Anterior Cingulate Cortex (the 'salience network' responsible for regulating attention between internal and external events, Craig, 2009) was also stronger in men, suggesting that they may be more likely to switch from an internal to external orientation in response to threat. Whittle, Yucel, Yap, and Allen (2011) concluded that the greater integration of emotional and cognitive circuitry in men means that they are more successful at regulating affective response to emotional stimuli. This contrasts with several studies that have testified to women's slower habituation to negative stimuli, suggesting a more protracted engagement with negative emotional stimuli (Andreano et al., 2014; Thomas et al., 2001; Williams et al., 2005). In support of this, Engman, Linnman, Van Dijk, and Milad (2016) confirmed greater resting state connectivity in men between the basolateral amygdala and ventromedial prefrontal cortex (implicated in emotional regulation). In short, this evidence tentatively suggests that emotional stimuli elicit responses in females that do not decline in strength as quickly as they do in males.

Sex differences in regional activation to emotional faces may be mediated by steroid hormones or neuropeptides, receptors for which are plentiful in the amygdala. The short-term effects of testosterone administration have been examined in a series of studies (reviewed in Bos, Panksepp, Blithe, & van Honk, 2012). Testosterone reduces fear as indexed by psychophysiological measures such as skin conductance response and fear potentiated startle (a physiological reflexive reaction to fear inducing stimulus), as well as behavioural measures such as reduced vigilance for fearful faces (van Honk, Peper, & Schutter, 2005). On this basis, it might be expected that testosterone would decrease amygdala reactivity to threat. In fact, testosterone selectively enhanced women's amygdala reactivity to angry faces (Hermans, Ramsey, & van Honk, 2008). This seemingly anomalous result was interpreted as enhanced preparedness for aggression rather than enhanced fear

(Hermans et al., 2008; Montoya, Terburg, Bos, & van Honk, 2012). But caution is warranted in interpreting these findings. Studies have found that testosterone enhancement of amygdala activation is not confined to threat expressions in either men (Derntl et al., 2009; Goetz et al., 2014) or women (Bos, van Honk, Ramsey, Stein, & Hermans, 2013; van Wingen et al., 2009), so it probably has a more general effect such as registering 'salience' rather than preparing for aggression. Furthermore, interpreting testosterone's effects in women is made more difficult because, unlike men's brains, women's brains have not been prenatally organised by testosterone exposure and indeed are protected from the masculinising effects of estrogens in utero by alpha-fetoprotein. The same hormone can have a markedly different, even opposite, effect on men and women (e.g. Thompson, Gupta, Miller, Mills, & Orr, 2004). In addition, the typical experimental dosage of 0.5 mg of testosterone raises a woman's testosterone level tenfold, well outside the normal female physiologic range: responses to this dosage therefore tell us little about the role of testosterone changes over normal ranges in women. In this review, the role of hormones such as testosterone will not be considered in depth (as, detailed in the final chapter, this domain is worthy of a volume in its own right) and will only be drawn upon occasionally.

The sex difference in amygdala response to frightening stimuli might underpin sex differences in self-reported susceptibility to fear as follows. The self-report method requires respondents to scan their autobiographical memory for episodes in which they felt afraid and, based on these recollections, to estimate the intensity or frequency of their fear. Women show more extensive, accurate and detailed autobiographical emotional memory than men which has been attributed to greater affective intensity at encoding and more frequent elaborative rehearsal (Andreano & Cahill, 2009). Emotional episodes tend to be well remembered: the encoding and consolidation of emotional memories are mediated by the basolateral amygdala's projections to the hippocampus and caudate nucleus, critical for visual and spatial memory (McGaugh, 2004). Women show greater resting connectivity than men between the amygdala and hippocampus (Kogler et al., 2016). Amygdala activation at the time of memory encoding correlates highly with later recall. This correlation is larger for stimuli that are rated as more emotionally intense. Women's stronger amygdala response to fear-inducing stimuli at the time of exposure would be expected to result in improved memory consolidation and hence greater availability at later recall. Ultimately this could enhance women's self-report of the intensity and frequency of subjective fear.

While the evidence reviewed thus far suggests that the amygdala may be in part responsible for stronger activation in women in response to fear-related stimuli, research also points to potential sex differences in the functioning of another key cortical area: the Orbitofrontal Cortex. Although Orbitofrontal Cortex activation continues to be popularly interpreted as reflecting inhibitory control, other interpretations have assigned it roles in decision making in terms of somatic markers (emotion-linked 'feelings'), representing stimulus-outcome relationships, assessing value assigned to differing outcomes, or tracking prediction errors (reviewed by Stalnaker, Cooch, & Schoenbaum, 2015). One suggestion that encompasses many of these functions is that the Orbitofrontal Cortex is a cognitive map of current workspace (Wilson, Takahashi, Schoenbaum, & Niv, 2014) by assimilating information from current sensory/perceptual inputs with memory data on previous events and responses. A recent study found that Orbitofrontal Cortex activity tracked 'unobservable' task states (i.e. relevant but unobservable decision relevant information e.g. information from working memory): States relevant to decision making but which are difficult to distinguish based on sensory data alone (Schuck, Ming, Wilson, & Niv, 2016). Hence the Orbitofrontal Cortex is especially important when behaviour is guided by mental simulations of future outcomes that have not been previously experienced. In the case of fear, these outcome simulations are likely to be their value (danger) and their implications for responses (avoidance). Women's stronger Orbitofrontal Cortex activity may indicate greater engagement with the aversive future consequences of an unfolding threat. The strength of this system can be disrupted by testosterone which reduces functional coupling between the Orbitofrontal Cortex and amygdala in women, consequently reducing the Orbitofrontal Cortex's regulatory functions and contributing to amygdala disinhibition (van Wingen, Mattern, Verkes, Buitelaar, & Fernandez, 2010). Longitudinally, rising levels of testosterone during male puberty are associated with reduced Orbitofrontal Cortex-amygdala connectivity (Peters, Jolles, Van Duijvenvoorde, Crone, & Peper, 2015; Spielberg et al., 2015) and reduced connectivity is also found in relatively 'fearless' psychopathic youths (Marsh et al., 2011). The activity of the Orbitofrontal Cortex and its connectivity with the amygdala may therefore be a key difference between men and women in the face of potentially fearful stimuli.

There is also consistent evidence of stronger Anterior Cingulate Cortex activation in women. The Anterior Cingulate Cortex has been implicated in the control of attention (Critchley et al. 2004), particularly under conditions of threat or danger which, as in the case of fear, involve uncertainty signalled by disrupted homeostasis and raised autonomic nervous system activity (Gasquoine, 2013). These situations require increased cognitive resources. Heightened Anterior Cingulate Cortex activation in women may indicate a decision to allocate greater processing resources to threat.

In conclusion, there is considerable evidence for sex differences in the *strength* of response of the fear system. Increased activation of key structures in the fear system (amygdala, Orbitofrontal Cortex and Anterior Cingulate Cortex) suggest a more robust fear response in women and one which is better integrated, given the greater connectivity between subcortical, frontal and sensory networks in women (Tunc et al., 2016, see also McClure et al., 2004). The impact of this more intense fear response is felt in a stronger motivation to avoid or escape fear-inducing situations when they are encountered. Behaviourally, the response to threat shown by men and women is one of degree rather than kind which challenges an alternative theory regarding sex differences ("Tend and Befriend": Taylor et al., 2000) and is discussed in detail in Chap. 7). Decisions to flee, freeze or fight depend on the proximity and escapability of the predator or hostile conspecific (Blanchard, Hynd, Minke, Minemoto, & Blanchard, 2001). In scenario studies in which respondents rated their likely response to situations of aggressive threat, there were significant correlations between the two sexes in their choice of behavioural response (Blanchard et al., 2001; Perkins & Corr, 2006). However, women rated their own fear and the danger of the situation higher than men and were more likely to select a defensive orientation away from the threat (running, hiding, freezing rather than attacking) and a stronger defensive intensity (Perkins & Corr, 2006). Prospectively greater fear during decision making, deters involvement in potentially dangerous activities by assigning a greater negative weight in the calculation of projected costs (see Chap. 6). Fear induced desistance would have enhanced survival in women encountering evolutionary challenges that would otherwise have led to costly long-term fitness consequences.

References

Andreano, J. M., & Cahill, L. (2009). Sex influences on the neurobiology of learning and memory. *Learning and Memory, 16*(4), 248–266.

Andreano, J. M., Dickerson, B. C., & Barrett, L. F. (2014). Sex differences in the persistence of the amygdala response to negative material. *Social Cognitive and Affective Neuroscience, 9*, 1388–1394.

Blanchard, D. C., Hynd, A. L., Minke, K. A., Minemoto, T., & Blanchard, R. J. (2001). Human defensive behaviors to threat scenarios show parallels to fear and anxiety-related defense patterns of non-human mammals. *Neuroscience and Biobehavioral Reviews, 25*, 761–770.

Bos, P. A., Panksepp, J., Blithe, R.-M., & van Honk, J. (2012). Acute effects of steroid hormones and neuropeptides on human social–emotional behavior: A review of single administration studies. *Frontiers in Neuroendocrinology, 33*, 17–35.

Bos, P. A., van Honk, J., Ramsey, N. F., Stein, D. J., & Hermans, E. J. (2013). Testosterone administration in women increases amygdala responses to fearful and happy faces. *Psychoneuroendocrinology, 38*, 808–817.

Craig, A. D. (2009). How do you feel–now? The anterior insula and human awareness. *Nature Reviews Neuroscience, 10*, 59–70.

Critchley, H. D., Wiens, S., Rotshtein, P., Ohman, A., & Dolan, R. J. (2004). Neural systems supporting interoceptive awareness. *Nature Neuroscience, 7*(2), 189–195.

Davis, F. C., Somerville, L. H., Ruberry, E. J., Berry, A. B. L., Shin, L. M., & Whalen, P. J. (2011). A tale of two negatives: Differential memory modulation by threat-related facial expressions. *Emotion, 11*(3), 647–655.

Derntl, B., Windischberger, C., Robinson, S., Kryspin-Exner, I., Gur, R. C., Moser, E., & Habel, U. (2009). Amygdala activity to fear and anger in healthy young males is associated with testosterone. *Psychoneuroendocrinology, 34*, 687–693.

Engman, J., Linnman, C., Van Dijk, K. R. A., & Milad, M. R. (2016). Amygdala subnuclei resting-state functional connectivity sex and estrogen differences. *Psychoneuroendocrinology, 63*, 34–42.

Etkin, A., Egner, T., & Kalisch, R. (2011). Emotional processing in anterior cingulate and medial prefrontal cortex. *Trends in Cognitive Sciences, 15*, 85–93.

Gasquoine, P. G. (2013). Localization of function in anterior cingulate cortex: From psychosurgery to functional neuroimaging. *Neuroscience and Biobehavioral Reviews, 37*, 340–348.

Goetz, S. M. M., Tang, L., Thomason, M. E., Diamond, M. P., Hariri, A. R., & Carré, J. M. (2014). Testosterone rapidly increases neural reactivity to threat in healthy men: A novel two-step pharmacological challenge paradigm. *Biological Psychiatry, 76*, 324–331.

Hammer, J. L., & Marsh, A. A. (2015). Why do fearful facial expressions elicit behavioral approach? Evidence from a combined approach-avoidance implicit association test. *Emotion, 15*(2), 223–231.

Hermans, E. J., Ramsey, N. F., & van Honk, J. (2008). Exogenous testosterone enhances responsiveness to social threat in the neural circuitry of social aggression in humans. *Biological Psychiatry, 63*, 263–270.

Kogler, L., Mueller, V. I., Seidel, E. M., Boubela, R., Kalcher, K., Moser, E., . . . Derntl, B. (2016). Sex differences in the functional connectivity of the amygdalae in association with cortisol. *NeuroImage, 134*, 410–423.

Lungu, O., Potvin, S., Tikasz, A., & Mendrek, A. (2015). Sex differences in effective fronto-limbic connectivity during negative emotion processing. *Psychoneuroendocrinology, 62*, 180–188.

Marsh, A. A., & Ambady, N. (2007). The influence of the fear facial expression on prosocial responding. *Cognition and Emotion, 21*(2), 225–247.

Marsh, A. A., Ambady, N., & Kleck, R. E. (2005). The effects of fear and anger facial expressions on approach- and avoidance-related behaviors. *Emotion, 5*, 119–124.

Marsh, A. A., Finger, E. C., Fowler, K. A., Jurkowitza, I. T. N., Schechter, J. C., Yua, H. H., . . . Blair, R. J. R. (2011). Reduced amygdala-orbitofrontal connectivity during moral judgments in youths with disruptive behavior disorders and psychopathic traits. *Psychiatry Research, 194*(3), 279–286.

McClure, E. B., Monk, C. S., Nelson, E. E., Zarahn, E., Leibenluft, E., Bilder, R. M., . . . Pine, D. S. (2004). A developmental examination of gender differences in brain engagement during evaluation of threat. *Biological Psychiatry, 55*, 1047–1055.

McGaugh, J. L. (2004). The amygdala modulates the consolidation of memories of emotionally arousing experiences. *Annual Review of Neuroscience, 27*, 1–28.

Montoya, E. R., Terburg, D., Bos, P. A., & van Honk, J. (2012). Testosterone, cortisol, and serotonin as key regulators of social aggression: A review and theoretical perspective. *Motivation and Emotion, 36*(1), 65–73.

Moriguchi, Y., Touroutoglou, A., Dickerson, B. C., & Barrett, L. F. (2014). Sex differences in the neural correlates of affective experience. *Social Cognitive and Affective Neuroscience, 9*, 591–600.

Perkins, A. M., & Corr, P. J. (2006). Reactions to threat and personality: Psychometric differentiation of intensity and direction dimensions of human defensive behavior. *Behavioural Brain Research, 169*, 21–28.

Peters, S., Jolles, D. J., van Duijvenvoorde, A. C. K., Crone, E. A., & Peper, J. S. (2015). The link between testosterone and amygdala-orbitofrontal cortex connectivity in adolescent alcohol use. *Psychonueroendocrinology, 53*, 117–126.

Schuck, N. W., Ming, B. C., Wilson, R. C., & Niv, Y. (2016). Human orbitofrontal cortex represents a cognitive map of state space. *Neuron, 91*, 1402–1412.

Sergerie, K., Chochol, C., & Armony, J. L. (2008). The role of the amygdala in emotional processing: A quantitative meta-analysis of functional neuroimaging studies. *Neuroscience and Biobehavioral Reviews, 32*, 811–830.

Spielberg, J. M., Forbes, E. E., Ladouceur, C. D., Worthman, C. M., Olino, T. M., Ryan, N. D., & Dahl, R. E. (2015). Pubertal testosterone influences threat-related amygdala–orbitofrontal cortex coupling. *Social Cognitive and Affective Neuroscience, 10*, 408–410.

Stalnaker, T. A., Cooch, N. K., & Schoenbaum, G. (2015). What the orbitofrontal cortex does not do. *Nature Neuroscience, 18*(5), 620–627.

Stevens, J. S., & Hamann, S. (2012). Sex differences in brain activation to emotional stimuli: A meta-analysis of neuroimaging studies. *Neuropsychologia, 50*, 1578–1593.

Taylor, S. E., Klein, L. C., Lewis, B. P., Gruenewald, T. L., Gurung, R. A. R., & Updegraff, J. A. (2000). Biobehavioral responses to stress in females: Tend-and-befriend, not fight-or-flight. *Psychological Review, 107*(3), 411–429.

Thomas, K. M., Drevets, W. C., Whalen, P. J., Eccard, C. H., Dahl, R. E., Ryan, N. D., & Casey, B. J. (2001). Amygdala response to facial expressions in children and adults. *Biological Psychiatry, 49*, 309–316.

Thompson, R., Gupta, S., Miller, K., Mills, S., & Orr, S. (2004). The effects of vasopressin on human facial responses related to social communication. *Psychoneuroendocrinology, 29*, 35–48.

Tunc, B., Solmaz, B., Parker, D., Satterthwaite, T. D., Elliott, M. A., Calkins, M. E., ... Verma, R. (2016). Establishing a link between sex-related differences in the structural connectome and behaviour. *Philosophical Transactions of the Royal Society B, 371*, 20150111.

van Honk, J., Peper, J. S., & Schutter, D. J. (2005). Testosterone reduces unconscious fear but not consciously experienced anxiety: Implications for the disorders of fear and anxiety. *Biological Psychiatry, 58*, 218–225.

van Wingen, G., Mattern, C., Verkes, R. J., Buitelaar, J., & Fernandez, G. (2010). Testosterone reduces amygdala–orbitofrontal cortex coupling. *Psychoneuroendocrinology, 35*, 105–113.

van Wingen, G. A., Zylicz, S. A., Pieters, S., Mattern, C., Verkes, R. J., Buitelaar, J. K., & Fernandez, G. (2009). Testosterone increases amygdala reactivity in middle-aged women to a young adulthood level. *Neuropsychopharmacology, 34*, 539–547.

Wager, T. D., Phan, K. L., Liberzon, I., & Taylor, S. F. (2003). Valence, gender, and lateralization of functional brain anatomy in emotion: A meta-analysis of findings from neuroimaging. *NeuroImage, 19*, 513–531.

Whittle, S., Yucel, M., Yap, M. B. H., & Allen, N. B. (2011). Sex differences in the neural correlates of emotion: Evidence from neuroimaging. *Biological Psychology, 87*, 319–333.

Williams, L. M., Barton, M. J., Kemp, A. H., Liddell, B. J., Peduto, A., Gordon, E., & Bryant, R. A. (2005). Distinct amygdala-autonomic arousal profiles in response to fear signals in healthy males and females. *NeuroImage, 28*(3), 618–625.

Wilson, R. C., Takahashi, Y. K., Schoenbaum, G., & Niv, Y. (2014). Orbitofrontal cortex as a cognitive map of task space. *Neuron, 81*, 267–279.

Wood, A., Rychlowska, M., Korb, S., & Niedenthal, P. (2016). Fashioning the face: Sensorimotor simulation contributes to facial expression recognition. *Trends in Cognitive Sciences, 20*, 227–240.

Chapter 5
Sex Differences in Time Course of Fear Response

Women may differ from men in the time course of their fear experience. Acquired and conditioned fears may persist for longer and each fear experience may extend for a longer duration. This creates a higher baseline of fear onto which situational fear cues may 'add' or 'stack'. These possibilities will be examined by reviewing sex differences in the areas of fear conditioning, sustained and phasic fear, potentiated (induced) fear, and more extended processing time for fear-inducing stimuli/events. If our hypothesis is correct, we would expect to find evidence of greater persistence of the experience of fear in women compared to men.

5.1 Evidence from Classical Conditioning Studies

Fear conditioning studies employ classical conditioning techniques to pair neutral cues (a conditioned stimuli, hereby referred to as CS) with fear-inducing aversive stimuli (an unconditioned stimuli, hereby referred to as UCS). A UCS that spontaneously evokes fear, such as an electric shock, is repeatedly paired with a neutral stimulus (CS+), such as a colour or letter sequence. Another neutral stimulus (CS-) is paired with the absence of shock. The responses to the CS+ and CS- are then measured. The differences in responses to the CS+ and CS- is taken as an index of fear conditioning. Following the acquisition procedure, extinction trials, extinction memory trials and reinstatement trials may be introduced. These are designed to determine how long the response to the CS+ takes to decay and can provide insight into how we acquire and process fearful stimuli.

The proposal that the higher prevalence of anxiety and post-traumatic stress disorder (PTSD) in women may be a result of a sex difference in the speed or strength of fear acquisition or its extinction (Lebron-Milad & Milad, 2012; Maeng & Milad, 2015) has prompted a substantial body of research in this area. There are two ways in which time course could provide evidence to support sex differences in fear. Faster conditioning among women in response to previously neutral or low-threat

A. Campbell et al., *Sex Differences in Fear Response*, SpringerBriefs in Anthropology, https://doi.org/10.1007/978-3-030-65280-7_5

stimuli would result in a wider range of fear-provoking stimuli. Alternatively, a longer latency (or failure) to extinguish conditioned fear in women would result in the long-term maintenance of fears that are realistically unlikely to signal danger, as seen in phobias. Stronger evidence has been found for the latter proposal than the former (see Chap. 3 for evidence surrounding phobias).

The bulk of conditioning research has been performed on rodents and sex differences in fear acquisition have been inconsistent (reviewed in Lebron-Milad & Milad, 2012; Lebron-Milad et al., 2012). Sex differences are found more reliably in the extinction phase with females' fear responses extinguishing more slowly (Baran, Armstrong, Niren, Hanna, & Conrad, 2009).

In humans, the most commonly used measure of conditioning is differential skin conductance response (SCR) to the CS+ versus CS-. Most studies report null findings (Kelly & Forsyth, 2007; Lebron-Milad et al., 2012; Rosenbaum et al., 2015; Zorawski, Cook, Kuhn, & LaBar, 2005) and some a larger differential SCR response in men (Milad et al., 2006, 2010). As with animal research. There appears to be no consistent sex difference in this domain. This however could be down to methodological limitations within this field (see Sect. 7.4.1).

With the advent of scanning technology, human studies have focused on neural correlates of fear conditioning (see Appendix B for a summary of relevant neuropsychological regions). During the acquisition of a conditioned response, sensory and cortical CS inputs arrive at the lateral amygdala where they are joined by UCS inputs. When they converge, CS inputs are potentiated via changes to synapses through the activity of the postsynaptic neurotransmitter receptors for N-methyl-D-aspartate (NMDA). The lateral amygdala connects to the amygdala's basal and central regions. The basal amygdala contains two populations of neurons differentially associated with fear expression and fear extinction (DeJean et al., 2015; Herry et al., 2008). The inhibition of fear during extinction depends on bidirectional connectivity between the basal amygdala and the ventromedial prefrontal cortex (corresponding to the rodent infralimbic IL area) which modulates amygdala activity. The basal amygdala also has reciprocal connections with the ventral hippocampus (which stores information about the context in which the learning took place) and the dorsal anterior cingulate cortex (corresponding to the rodent prelimbic cortex). The function of the dorsal anterior cingulate cortex includes detection of salience, allocation of attention (activity increases as a function of UCS unexpectedness), and behavioural and autonomic fear responses (Milad et al., 2007). The central amygdala communicates with the hypothalamus, generating autonomic fear responses, and with the periaqueductal grey subserving behavioural fear expression. The insula has also been implicated in fear conditioning (Klucken et al., 2009) and may be responsible for integrating interoceptive cues of arousal or pain with the emotional experience of fear (Craig, 2009; Damasio et al., 2000).

The above suggests several key areas that may be critical in maintaining the fear system. However, few imaging studies have directly examined sex differences within them. Using painful rectal distension as a UCS, Benson et al. (2014) found that women displayed significantly greater insula activity in the CS+/CS- contrast during late acquisition, reflecting enhanced visceral sensitivity to anticipated pain, although there were no sex differences in self-reported pain or stress. During extinction, women showed reduced activation in the posterior cingulate, a region implicated in pain modulation, suggesting that women may be less efficient at re-learning stimulus properties (that CS+ no longer predicts the painful UCS). During the reinstatement phase, women showed stronger activation than men in areas associated with memory consolidation and retrieval (hippocampus, thalamus and cerebellum) suggesting that previously extinguished fears were more easily reactivated. This study would suggest that the initial conditioned fear of pain persisted for longer in women and took longer to reverse.

Women were also more reactive to fear conditioning when an electric shock was used as the UCS (Lebron-Milad et al., 2012). Although there was no sex difference in skin conductance responses in any of the conditioning phases, women showed greater activity than men in a number of brain regions. During acquisition, women showed increased activation in the amygdala, and the dorsal anterior cingulate cortex associated with fear enhancement and salience (Etkin, Egner, Peraza, Kandel, & Hirsch, 2006). During the extinction phase, women showed more activation in the dorsal anterior cingulate cortex, medial prefrontal cortex and left hypothalamus, and during extinction recall in the insula. Two further studies used a shock conditioning paradigm designed to prevent conscious awareness of the association between the CS and the UCS (Merz et al., 2010, 2012). No sex differences in differential activation to CS+/CS- were found in regions of interest, such as the amygdala, hippocampus, thalamus, insula, frontal cortex, or anterior hippocampal gyri. Taken together, these studies are suggestive of the greater importance of conscious, anticipatory awareness of threat in women compared to men.

Women's stronger neural response to fear acquisition and weaker extinction has prompted consideration of the role of estrogen in the process of acquiring and maintaining fear (Stockhorst & Antov, 2016). Estradiol (an estrogen steroid hormone) administration enhances the consolidation of extinction memory (Graham & Milad, 2013; Milad et al., 2010) and higher estradiol levels during extinction learning improves later extinction recall (i.e. recall of a safety memory). Results suggest that women with high levels of estradiol performed similarly to men and significantly better than women whose extinction training took place during their low estradiol phase (Milad et al., 2010). Women in the high-estradiol phase also had increased ventromedial prefrontal cortex activation (which is associated with greater fear reduction) during extinction learning and recall which was significantly correlated with their estrogen levels (Zeidan et al., 2011). In short, this suggests that low levels of estradiol increases the persistence of fear related memory, while rising levels lead to greater fear reduction. We will return to the potential significance of this relationship in Sect. 7.4.2.

Stress—and its associated hormone cortisol—have also been examined to address whether women's typically greater anxiety enhances their acquisition of learned fears. For autonomic measures, studies have indicated a sex-specific effect, replicating rodent findings (e.g. Wood & Shors, 1998). Cortisol enhanced the acquisition of differentiated skin conductance responses in men but not women (Jackson, Payne, Nadel, & Jacobs, 2006; Zorawski et al., 2005; Zorawski, Blanding, Kuhn, & LaBar,

2006). However, this effect was reversed at a neuropsychological level (Stark et al., 2006). For women, cortisol enhanced differential CS+/CS- neural reactivity in the prefrontal regions associated with fear acquisition (anterior cingulate, lateral occipital frontal cortex and medial prefrontal cortex), while reducing activation in these same regions in men. Using implicit (unaware) fear conditioning, Merz et al. (2010) found similar sexually-dimorphic effects at the subcortical level. Relative to a placebo, cortisol increased differential insula activity to CS+/CS- in women while decreasing it in men, with a similar trend in the hippocampus and thalamus. Similar results were found when stress was induced behaviourally, rather than hormonally (Merz et al., 2013). Although at the autonomic level stress does not improve women's fear learning, it strengthens activity in the associated neural circuitry. Women's typically greater anxiety appears to strengthen the neural representations of acquired fears.

At this point, it is worth summarising what is a complex combination of studies on fear conditioning, and how this pertains to the hypothesis that women have an increased time course for fear. The evidence thus far seems to suggest that women extinguish learned fears more slowly than men (Maeng & Milad, 2015), despite the absence of consistent sex differences in the acquisition of differential skin conductance responses during conditioning. When the CS-UCS association is severed during extinction, women have greater difficulty in relearning that the stimulus is now safe, and continue to display emotional and autonomic reactivity to the now-neutral stimuli. This sex difference in extinction appears to be hormonally mediated by estradiol levels. Women who underwent extinction training in their high estradiol phase had extinction recall that was comparable to men and significantly better than women who were trained during their low estrogen phase. Estradiol levels were also positively correlated with activation of the ventromedial prefrontal cortex, a structure strongly implicated in reducing fear. Stress disrupts neural correlates of men's fear conditioning but enhances women's, suggesting that the inherent stress associated with confronting threat improves women's ability to form crucial associations between contextual cues and danger. Stress also increases women's, but not men's, attentional bias toward threatening stimuli (Carr, Scully, Webb, & Felmingham, 2016). Taken in tandem, these findings would suggest that the impact of fear in women is likely to be more persistent than in men.

5.2 Evidence Beyond Classical Conditioning

While classical conditioning studies demonstrate the expected sex differences, women also show stronger conditioning than men in studies using operant and observational paradigms as well. The operant concept of negative reinforcement underlies the behavioural avoidance paradigm in which participants learn to make an avoidant response when a cue signals the arrival of a threatening stimulus. This was adapted for human research using a computerised game in which the presence of a cue signalled the imminent arrival of a bomb associated with explosions and point loss (Sheynin et al., 2014). Participants were instructed to acquire as many points as possible while 'hiding' from the bomb. The duration of women's avoidance (operationalized as hiding time in response to the cue) was significantly longer than men's. This was replicated in a further study (Sheynin, Beck, Servatius, & Myers, 2014) and the effect persisted into extinction trials. Sex differences in observational learning of panic were examined in a study by Kelly and Forsyth (2007) in which participants watched a confederate breathing CO_2 enriched air (which is a method of experimentally inducing fear in the recipient). The CS+ was paired with images of the confederate displaying panic symptoms, while CS- was paired with normal breathing. During vicarious conditioning, skin conductance response measures did not differ by sex, but women reported greater distress and dislike of the CS+ than men and, during extinction trials, women maintained a stronger differential response to CS+ and CS- on these measures.

In recent years, rodent and human research has supported a distinction between phasic (acute) fear and ***sustained*** fear. Phasic fear is a temporary and acute reactive state that motivates avoidance and escape, after which the emotion quickly dissipates (Epstein, 1972; Gray & McNaughton, 2000). The source of danger is present, specific and immediate (Adolphs, 2013; Lang, Davis, & Öhman, 2000). Sustained fear refers to anticipatory fear of a yet-to-be-encountered aversive stimulus. Sustained fear has sometimes been treated as synonymous with anxiety. Although there is overlap between structures involved in the fear and anxiety systems (Tovote, Fadok, & Luthi, 2015), there is evidence for their independence in terms of the engagement of different amygdala subnuclei, the role of peptide hormones, and a differential response to pharmacological interventions (Davis, Walker, Miles, & Grillon, 2010; Tovote et al., 2015).

Anxiety is a response to a stimulus which is ambiguous or uncertain in terms of threat. The temporal orientation is to the indeterminate future. It is associated with a protracted state of generalised arousal that can persist even when the individual is removed from the triggering situation. By contrast, sustained fear is operationalized in terms of the local and temporal proximity of the aversive stimulus—the uncertainty surrounds not 'if' but 'when' the danger will occur (Davis et al., 2010). For this reason, in the following discussion, sustained fear will be treated as a state of anticipatory fear rather than anxiety. Being anticipatory rather than reactive, a stronger sustained fear system in women could provide an advantage by extending baseline fear with a more sustained readiness to respond to a specific fear cue. The temporal duration of fear would be lengthened, thus potentially increasing the responses of the other hypothesised fear components i.e. salience, strength and conscious accessibility (awareness). Geared as it is to future danger, sustained fear is especially relevant to precautionary processing and preventative action (Hinds et al., 2010) which, in the context of this thesis would have obvious adaptive advantages for women in particular.

The 'extended amygdala' is the proposed site of sustained fear. In addition to the central nucleus of the amygdala, it includes the lateral bed nucleus of the stria terminalis (commonly referred to as BNST). Resting functional activity is coordinated between the two regions. They are composed of similar cell types and show a similar pattern of structural connectivity including projections to the hypothalamus and brain stem, and incoming input from the prefrontal and insular cortices (Fox, Oler, Tromp, Fudge, & Kalin, 2015). Based largely on rodent studies, Davis, Walker, and Lee (1997) proposed that the amygdala was implicated in phasic (acute) fear whereas the BNST was associated with sustained fear. When an initial threat persists in time, transient amygdala activation is superseded by BNST activation which maintains anticipatory hypervigilance (Alvarez, Chen, Bodurka, Kaplana, & Grillon, 2011). 'Sustained fear' describes this state of watchful risk assessment (Davis et al., 2010) and the BNST responds strongly to scenarios presenting a high risk of physical harm (Coaster et al., 2011).

Only a few human brain imaging studies have examined activity in the BNST because the region is not identified by automatic labelling software in imaging apparatus, and its location, shape and size make it hard to identify. Somerville, Whalen, and Kelley (2010) scanned participants as they viewed a line which fluctuated in height and which, if it exceeded a certain height threshold, signalled the later delivery of an electric shock. The ventral basal forebrain (containing the BNST) and the insula were increasingly activated as the line approached threshold, with anxious participants showing more strongly enhanced activation. BNST also showed activation when participants viewed a spider approaching their foot (Mobbs et al., 2010), when anticipating an electric shock associated with being captured by enemies in a video game (Mobbs et al., 2007), and when contemplating the possibility of serious physical injury (Coaster et al., 2011). Alvarez et al. (2011) examined activation under predictable shock (signalled by a tone) and unpredictable shock, with different virtual reality contexts serving as visual cues for conditions. In both conditions, there was transient dorsal amygdala activation, but only the unpredictable shock condition produced sustained activity in the BNST. Sustained fear was also associated with cortical areas implicated in maintaining hypervigilance (anterior insula and frontoparietal regions), and emotional regulation and stress (ventral amygdala-hippocampal regions and anterior cingulate).

The sexually dimorphic nature of the BNST has been known for over 25 years with a 2.5 times greater volume in men than women (Allen & Gorski, 1990). There are also sex differences in structural connectivity (Avery et al., 2014). Seventy-six per cent of regions connected to the BNST (including the amygdala bilaterally) show significantly greater connectivity in women than men. Startle modulation was used as a tool to examine human sex differences in phasic and sustained fear by Grillon (2008; Grillon et al., 2011). Participants performed in three conditions: predictable shocks (signalled by a CS cue), unpredictable shocks and no shocks. Acoustic startle stimuli delivered after the cue onset constituted tests of fear potentiation and no sex differences were found. Acoustic stimuli delivered in the inter-trial intervals constituted tests of sustained fear (i.e. fear of the experimental context in general rather than a signalled impending shock) and here women showed a stronger response than men. The same pattern of sex difference was found for self-rated anxiety. This sex difference has been replicated with children (Schmitz et al., 2011).

To summarise this evidence so far, studies of sustained fear suggest women may show increased vigilance in response to uncertainty when they anticipate the subsequent arrival of a feared stimulus. In terms of their implications for behavioural response, phasic and sustained fear cues can be viewed as situated on a continuum of immediacy or certainty, ranging from current certain danger to suspected future danger (Bolles & Fanselow, 1980). When danger is anticipated, the sustained fear system interrupts ongoing behaviour, instigates a search for additional information and assesses risk. But when the danger is immediate, a faster and more automatic response is required. Activation of the BNST during sustained fear is accompanied by activation in the anterior insula and frontal regions associated with vigilance and planning (Alvarez et al., 2011). Using a computer-simulated chase through a maze by a virtual predator with the ability to inflict pain when caught (via an electric shock to the participant), Mobbs et al. (2007, 2009) identified a shift from corticolimbic control (including the ventromedial prefrontal cortex, anterior cingulate cortex and hippocampus) when the threat was more remote, to periaqueductal grey control (associated with pain expectation, dread and reduced confidence of escape) when the attack actually occurred. Hence the relative latency and certainty of danger form a response continuum from context-sensitive anticipatory watchfulness to an evolutionarily prepared flight-or-fight response via the periaqueductal grey. If the more protracted time course associated with sustained fear was more available to women, it would permit the recruitment of frontal regions capable of formulating anticipatory and pre-emptive avoidance strategies. While phasic fear is typically short-lived and measured in seconds, sustained fear may last for several minutes. Given its anticipatory nature with the associated possibility of escape, a more protracted sustained fear system would be beneficial for women's and their children's survival.

5.3 Additional Evidence for Sex Differences in the Time Course of Fear

Evidence of a sex difference in the time course of fear has also been found in studies where this was not the originally theorised focus of the research. Studies of fear potentiation support a sex difference favouring woman in the anticipation of fear. Recall that this paradigm involves the instigation of sustained low-level 'background' fear in order to examine its effect on acute fear reactions. It thus corresponds to a situation of sustained fear culminating in an episode of phasic fear. In these studies, participants are presented with a sudden-onset auditory tone while viewing fear-inducing images or under threat of shock. Facial startle, recorded by electrodes above the orbicularis oculi muscle (muscles controlling the eyelids), is compared with baseline startle or startle during exposure to neutral images to quantify the increase in muscle contraction attributable to the fear-inducing stimuli. The neural circuitry of potentiated startle occurs within milliseconds. Acoustic pathways carry auditory information to the caudal pontine reticular nucleus where it meets afferent (incoming) information from the amygdala. Damage to this afferent pathway eliminates fear modulation of the response while leaving baseline startle intact. An advantage of the facial startle measure over other physiological measures such as heart rate or skin conductance is that the startle response is specifically sensitive to valence (fear) rather than arousal in general (Davis et al., 2010) and it is well-conserved across mammalian species (Lang et al., 2000).

As reviewed earlier, most studies find no sex differences in baseline startle. Yet the priming effect of extended anticipation on subsequent fear is stronger in women. Under fear potentiation, stronger startle responses are more consistently found in women (Bradley, Codispoti, Sabatinelli, & Lang, 2001; Lang, Greenwald, Bradley, & Hamm, 1993). A more extreme potentiated startle is associated with facing an approaching and uncontrollable threat, while an avoidable threat results in a less marked response (Löw, Weymar, & Hamm, 2015), suggesting that women may experience the same threat as more dangerous than men. A cross-sectional study of adults and children (aged 3–9 years) found that tones delivered during emotionally-valenced films induced a higher overall startle magnitude and a larger probability of eliciting a response in females (Quevedo, Smith, Donzella, Schunk, & Gunnar, 2010). In a longitudinal study of 10–13-year-old adolescents, girls showed stronger fear potentiation at all pubertal stages and fear potentiation increased with pubertal stage in girls but not boys (suggesting this strengthens as girls get closer to reproductive onset). In young adulthood, women showed stronger potentiated startle when viewing fearful and angry faces than men (Anokhin & Golosheykin, 2010). When exposed to emotional slides, women demonstrated greater overall amplitude of startle reflex which was larger for unpleasant than pleasant or neutral images, whereas men's response did not differ by slide valence (Bianchin & Angrilli, 2012). These studies therefore suggest that anticipated fear has a more pronounced effect in women that appears to strengthen as they approach reproductive maturity.

In a direct examination of sex differences in the time course of fear potentiation, men's and women's affective ratings and the magnitude of fear potentiated startle were measured during presentation and after the offset of pleasant, neutral and negative images (Gard & Kring, 2007). As expected, women scored higher on a trait measure of aversive motivational system strength and rated negative images as less pleasant and more arousing. Both sexes showed fear potentiation in the presence of negative images but, unlike men, women continued to show this fear enhancement during the recovery period when the images were no longer visible. This again suggests that women experience fear over a longer time course than men.

The hormone testosterone has been examined as a mediator of this sex difference. A placebo-controlled within-subjects study found that administration of testosterone to women reduced their fear-potentiated startle in response to threatened electric shock while leaving baseline startle unaffected (Hermans, Putman, Baas, Koppeschaar, & van Honk, 2006). However, a follow-up study replicated this moderating effect of testosterone only in a subgroup of highly anxious women (Hermans et al., 2007). For a more general review of the role of testosterone in motivating adaptive responses to competition and threat, see Archer (2006).

Studies of ***attentional bias*** strongly suggest different time courses of fear in men and women (but see Chap. 6 for discussion of how these provide scant support for sex differences in sensitivity to fear-provoking stimuli). Electroencephalogram

(EEG) studies have been conducted using the emotional Stroop test (measuring reaction times to threat-related compared to neutral words; see Appendix A) to measure changes to event-related potentials (brain wave responses to a specific stimuli or cognition). There are many measured event-related potentials in EEG studies but one of the most common is referred to as P300. P300 is an elicited event-related potential associated with an individual's reaction to the presentation of a stimulus (rather than any specific attribute of the stimulus itself). Sass et al. (2010) found longer latencies in women at P300 following threat-related words, consistent with the idea that they spend more time processing threatening stimuli. This finding has been replicated by Pintzinger, Pfabigan, Tran, Kryspin-Exner, and Lamm (2016) who also found that women, but not men, showed a longer P300 latency specifically for threatening words, indicating more extended evaluation and elaborated processing. This evidence should be read in conjunction with studies reviewed earlier (see strength in Chap. 4) of women's more protracted neural processing of threatening stimuli. This involved not merely a failure to acclimatise to previously-seen threat faces and stimuli (Gard & Kring, 2007), but a pattern of increasing amygdala activation over time (Andreano, Dickerson, & Barrett, 2014; Thomas et al., 2001) even within a single study (Williams et al., 2005). Altogether, these studies suggest that women spend more time processing threatening stimuli.

Other tentative evidence that women's attention is captured for longer by threat comes from the dot-probe task (see Appendix A). Tan, Ma, Gao, Wu, and Fang (2011) examined attentional bias in this task and suggested that the 800 ms response window allows for the possibility that women (especially anxious women) may have difficulty disengaging attention from aversive stimuli. A sex difference in latency to disengage would indicate that women continue exploration of stimuli after they have been established as threatening, rather than detecting threat more swiftly then men. However, evidence for a sex difference in disengagement currently remains equivocal (Pintzinger et al., 2016; Tan et al., 2011; Tran, Lamplmayr, Pintzinger, & Pfabigan, 2013).

To conclude our discussion on time course, there is much evidence that women are slower to shed fear responses to cues that are not reliably associated with danger, they are more likely to maintain a state of watchful sustained fear when danger is suspected, their acute fear responses are augmented by low-level anticipatory fear, they process threatening stimuli for longer, and adjust more slowly to threat cues. This extended time course of fear means, for example, that women are more alert to danger as they enter potentially risky situations and are consequently more attentive for relevant cues. The longer latency before danger is actually encountered allows for the engagement of frontal regions in the brain associated with planning for evasive action. The culmination of these sex differences in the time course of fear would have particularly strong fitness consequences for women, particularly when they become the mothers of infants (Hahn-Holbrook, et al., 2011).

References

Adolphs, R. (2013). The biology of fear. *Current Biology, 23*, R79–R93.

Allen, L. S., & Gorski, R. A. (1990). Sex difference in the bed nucleus of the stria terminalis of the human brain. *Journal of Comparative Neurology, 302*(4), 697–706.

Alvarez, R. P., Chen, G., Bodurka, J., Kaplana, R., & Grillon, C. (2011). Phasic and sustained fear in humans elicits distinct patterns of brain activity. *NeuroImage, 55*(1), 389–400.

Andreano, J. M., Dickerson, B. C., & Barrett, L. F. (2014). Sex differences in the persistence of the amygdala response to negative material. *Social Cognitive and Affective Neuroscience, 9*, 1388–1394.

Anokhin, A. P., & Golosheykin, S. (2010). Startle modulation by affective faces. *Biological Psychology, 83*, 37–40.

Archer, J. (2006). Testosterone and human aggression: An evaluation of the challenge hypothesis. *Neuroscience and Biobehavioral Reviews, 30*(3), 319–345.

Avery, S. N., Clauss, J. A., Winder, D. G., Woodward, N., Heckers, S., & Blackford, J. U. (2014). BNST circuitry in humans. *NeuroImage, 91*, 311–323.

Baran, S. E., Armstrong, C. E., Niren, D. C., Hanna, J. J., & Conrad, C. D. (2009). Chronic stress and sex differences on the recall of fear conditioning and extinction. *Neurobiology of Learning and Memory, 91*, 323–332.

Benson, S., Kattoor, J., Kullmann, J. S., Hofmann, S., Engler, H., Forsting, M., ... Elsenbruch, S. (2014). Towards understanding sex differences in visceral pain: Enhanced reactivation of classically-conditioned fear in healthy women. *Neurobiology of Learning and Memory, 109*, 113–121.

Bianchin, M., & Angrilli, A. (2012). Gender differences in emotional responses: A psychophysiological study. *Physiology and Behavior, 105*, 925–932.

Bolles, R. C., & Fanselow, M. S. (1980). A perceptual-defensive-recuperative model of fear and pain. *Behavioral and Brain Sciences, 3*, 291–301.

Bradley, M. M., Codispoti, M., Sabatinelli, D., & Lang, P. J. (2001). Emotion and motivation II: Sex differences in picture processing. *Emotion, 1*(3), 300–319.

Carr, A. R., Scully, A., Webb, M., & Felmingham, K. L. (2016). Gender differences in salivary alpha-amylase and attentional bias towards negative facial expressions following acute stress induction. *Cognition and Emotion, 30*(2), 315–324.

Coaster, M., Rogers, B. P., Jones, O. D., Viscusi, W. K., Merkle, K. L., Zald, D. H., & Gore, J. C. (2011). Variables influencing the neural correlates of perceived risk of physical harm. *Cognitive, Affective and Behavioral Neuroscience, 11*, 494–507.

Craig, A. D. (2009). How do you feel–now? The anterior insula and human awareness. *Nature Reviews Neuroscience, 10*, 59–70.

Damasio, A. R., Grabowski, T. J., Bechara, A., Damasio, H., Ponto, L. L., Parvizi, J., & Hichwa, R. D. (2000). Subcortical and cortical brain activity during the feeling of self-generated emotions. *Nature Neuroscience, 3*, 1049–1056.

Davis, M., Walker, D. L., & Lee, Y. (1997). Amygdala and bed nucleus of the stria terminalis: Differential roles in fear and anxiety measured with the acoustic startle reflex. *Philosophical Transactions of the Royal Society of London. Series B: Biological Sciences, 352*, 1675–1687.

Davis, M., Walker, D. L., Miles, L., & Grillon, C. (2010). Phasic vs sustained fear in rats and humans: Role of the extended amygdala in fear vs anxiety. *Neuropsychopharmacology, 35*, 105–135.

Dejean, C., Courtin, J., Rozeske, R. R., Bonnet, M. C., Dousset, V., Michelet, T., & Herry, C. (2015). Neuronal circuits for fear expression and recovery: Recent advances and potential therapeutic strategies. *Biological Psychiatry, 78*, 298–306.

Epstein, S. (1972). The nature of anxiety with emphasis upon its relationship to expectancy. In C. D. Spielberger (Ed.), *Anxiety: Current trends in theory and research* (Vol. 2, pp. 291–337). New York: Academic Press.

Etkin, A., Egner, T., Peraza, D. M., Kandel, E. R., & Hirsch, J. (2006). Resolving emotional conflict: A role for the rostral anterior cingulate cortex in modulating activity in the amygdala. *Neuron, 51*, 871–882.

Fox, A. S., Oler, J. A., Tromp, D. P. M., Fudge, J. L., & Kalin, N. H. (2015). Extending the amygdala in theories of threat processing. *Trends in Neurosciences, 38*(5), 319–329.

Gard, M. G., & Kring, A. M. (2007). Sex differences in the time course of emotion. *Emotion, 7*(2), 429–437.

Graham, B. M., & Milad, M. R. (2013). Blockade of estrogen by hormonal contraceptives impairs fear extinction in female rats and women. *Biological Psychiatry, 73*, 371–378.

Gray, J. A., & McNaughton, N. (2000). *The neuropsychology of anxiety* (2nd ed.). New York: Oxford University Press.

Grillon, C. (2008). Greater sustained anxiety but not phasic fear in women compared to men. *Emotion, 8*(3), 410–413.

Grillon, C., Heller, R., Hirschhorn, E., Kling, M. A., Pine, D. S., Schulkin, J., & Vythilingam, M. (2011). Acute hydrocortisone treatment increases anxiety but not fear in healthy volunteers: A fear-potentiated startle study. *Biological Psychiatry, 69*(6), 549–555.

Hahn-Holbrook, J., Holt-Lunstad, J., Holbrook, C., Coyne, S. M., & Lawson, E. T. (2011). Maternal defense: Breast feeding increases aggression by reducing stress. *Psychological Science, 22*(10), 1288–1295.

Hermans, E. J., Putman, P., Baas, J. M., Gecks, N. M., Kenemans, J. L., & van Honk, J. (2007). Exogenous testosterone attenuates the integrated central stress response in healthy young women. *Psychoneuroendocrinology, 32*, 1052–1061.

Hermans, E. J., Putman, P., Baas, J. M., Koppeschaar, H. P., & van Honk, J. (2006). A single administration of testosterone reduces fear-potentiated startle in humans. *Biological Psychiatry, 59*, 872–874.

Herry, C., Ciocchi, S., Senn, V., Demmou, L., Muller, C., & Luthi, A. (2008). Switching on and off fear by distinct neuronal circuits. *Nature, 454*, 600–606.

Hinds, A. L., Woody, E. Z., Drandic, A., Schmidt, L. A., Van Ameringen, M., Coroneos, M., & Szechtman, H. (2010). The psychology of potential threat: Properties of the security motivation system. *Biological Psychology, 85*, 331–337.

Jackson, E. D., Payne, J. D., Nadel, L., & Jacobs, W. J. (2006). Stress differentially modulates fear conditioning in healthy men and women. *Biological Psychiatry, 59*, 516–522.

Kelly, M. M., & Forsyth, J. P. (2007). Sex differences in response to an observational fear conditioning procedure. *Behavior Therapy, 38*, 340–349.

Klucken, T., Kagerer, S., Schweckendiek, J., Tabbert, K., Vaitl, D., & Stark, R. (2009). Neural, electrodermal and behavioral response patterns in contingency aware and unaware subjects during a picture-picture conditioning paradigm. *Neuroscience, 158*, 721–731.

Lang, P. J., Davis, M., & Öhman, A. (2000). Fear and anxiety: Animal models and human cognitive psychophysiology. *Journal of Affective Disorders, 61*(3), 137–159.

Lang, P. J., Greenwald, M. K., Bradley, M. M., & Hamm, A. O. (1993). Looking at pictures: Affective, facial, visceral, and behavioral reactions. *Psychophysiology, 30*(3), 261–273.

Lebron-Milad, K., Abbs, B., Milad, M. R., Linnman, C., Rougemount-Bücking, A., Zeidan, M. A., ... Goldstein, J. M. (2012). Sex differences in the neurobiology of fear conditioning and extinction: A preliminary fMRI study of shared sex differences with stress-arousal circuitry. *Biology of Mood and Anxiety Disorders, 2*, 7.

Lebron-Milad, K., & Milad, M. R. (2012). Sex differences, gonadal hormones and the fear extinction network: Implications for anxiety disorders. *Biology of Mood and Anxiety Disorders, 2*, 3.

Löw, A., Weymar, M., & Hamm, A. O. (2015). When threat is near, get out of here: Dynamics of defensive behavior during freezing and active avoidance. *Psychological Science, 26*(11), 1706–1716.

Maeng, L. Y., & Milad, M. R. (2015). Sex differences in anxiety disorders: Interactions between fear, stress, and gonadal hormones. *Hormones and Behavior, 76*, 106–117.

Merz, C. J., Tabbert, K., Schweckendiek, J., Klucken, T., Vaitl, D., Stark, R., & Wolf, O. T. (2010). Investigating the impact of sex and cortisol on implicit fear conditioning with fMRI. *Psychoneuroendocrinology, 35*, 33–46.

Merz, C. J., Tabbert, K., Schweckendiek, J., Klucken, T., Vaitl, D., Stark, R., & Wolf, O. T. (2012). Oral contraceptive usage alters the effects of cortisol on implicit fear learning. *Hormones and Behavior, 62*, 531–538.

Merz, C. J., Wolf, O. T., Schweckendiek, J., Klucken, T., Vaitl, D., & Stark, R. (2013). Stress differentially affects fear conditioning in men and women. *Psychoneuroendocrinology, 38*, 2529–2541.

Milad, M. R., Goldstein, J. M., Orr, S. P., Wedig, M. M., Klibanski, A., Pitman, R. K., & Rauch, S. L. (2006). Fear conditioning and extinction: Influence of sex and menstrual cycle in healthy humans. *Behavioral Neuroscience, 120*(6), 1196–1203.

Milad, M. R., Quirk, G. J., Pitman, R. K., Orr, S. P., Fischl, B., & Rauch, S. L. (2007). A role for the human dorsal anterior cingulate cortex in fear expression. *Biological Psychiatry, 62*, 1191–1194.

Milad, M. R., Zeidan, M. A., Contero, A., Pitman, R. K., Klibanski, A., Rauch, S. L., & Goldstein, J. M. (2010). The influence of gonadal hormones on conditioned fear extinction in healthy humans. *Neuroscience, 168*(3), 652–658.

Mobbs, D., Marchant, J. L., Hassabis, D., Seymour, B., Tan, G., Gray, M., … Frith, C. D. (2009). From threat to fear: The neural organization of defensive fear systems in humans. *Journal of Neuroscience, 29*, 12236–12243.

Mobbs, D., Petrovic, P., Marchant, J. L., Hassabis, D., Weiskopf, N., Seymour, B., … Frith, C. D. (2007). When fear is near: Threat imminence elicits prefrontal-periaqueductal gray shifts in humans. *Science, 317*, 1079–1083.

Mobbs, D., Yu, R., Rowe, J. B., Eich, H., FeldmanHall, O., & Dalgleish, T. (2010). Neural activity associated with monitoring the oscillating threat value of a tarantula. *Proceedings of the National Academy of Sciences U S A, 107*, 20582–20586.

Pintzinger, N. M., Pfabigan, D. M., Tran, U. S., Kryspin-Exner, I., & Lamm, C. (2016). Attentional biases in healthy adults: Exploring the impact of temperament and gender. *Journal of Behavior Therapy and Experimental Psychiatry, 52*, 29–37.

Quevedo, K., Smith, T., Donzella, B., Schunk, E., & Gunnar, M. (2010). The startle response: Developmental effects and a paradigm for children and adults. *Developmental Psychobiology, 52*, 78–89.

Rosenbaum, B. L., Bui, E., Marin, M.-F., Holt, D., Lasko, N. B., Pitman, R. K., … Milad, M. R. (2015). Demographic factors predict magnitude of conditioned fear. *International Journal of Psychophysiology, 98*(1), 59–64.

Sass, S. M., Heller, W., Stewart, J. L., Silton, R. L., Edgar, J. C., Fisher, J. E., … Miller, G. A. (2010). Time course of attentional bias in anxiety: Emotion and gender specificity. *Psychophysiology, 47*(2), 247–259.

Schmitz, A., Merikangas, K., Swendsen, H., Cui, L., Heaton, L., & Grillon, C. (2011). Measuring anxious responses to predictable and unpredictable threat in children and adolescents. *Journal of Experimental Child Psychology, 110*(2), 159–170.

Sheynin, J., Beck, K. D., Servatius, R. J., & Myers, C. E. (2014). Acquisition and extinction of human avoidance behavior: Attenuating effect of safety signals and associations with anxiety vulnerabilities. *Frontiers in Behavioral Neuroscience, 8*, 323.

Sheynin, J., Beck, K. D., Servatius, R. J., Myers, C. E., Pang, K. C. H., Servatius, R. J., … Myers, C. E. (2014). Behaviourally inhibited temperament and female sex, two vulnerability factors for anxiety disorders, facilitate conditioned avoidance (also) in humans. *Behavioural Processes, 103*, 228–235.

Somerville, L. H., Whalen, P. J., & Kelley, W. M. (2010). Human bed nucleus of the stria terminalis indexes hypervigilant threat monitoring. *Biological Psychiatry, 68*(5), 416–424.

Stark, R., Wolf, O. T., Tabbert, K., Kagerer, S., Zimmermann, M., Kirsch, P., & Vaitl, D. (2006). Influence of the stress hormone cortisol on fear conditioning in humans: Evidence for sex differences in the response of the prefrontal cortex. *NeuroImage, 32*(3), 1290–1298.

Stockhorst, U., & Antov, M. I. (2016). Modulation of fear extinction by stress, stress hormones and estradiol: A review. *Frontiers in Behavioral Neuroscience, 9*, 359.

Tan, J., Ma, Z., Gao, X., Wu, Y., & Fang, F. (2011). Gender difference of unconscious attentional bias in high trait anxiety individuals. *PLoS One, 6*(5), e20305.

Thomas, K. M., Drevets, W. C., Whalen, P. J., Eccard, C. H., Dahl, R. E., Ryan, N. D., & Casey, B. J. (2001). Amygdala response to facial expressions in children and adults. *Biological Psychiatry, 49*, 309–316.

Tovote, P., Fadok, J. P., & Luthi, A. (2015). Neuronal circuits for fear and anxiety. *Nature Reviews Neuroscience, 16*, 317–331.

Tran, U. S., Lamplmayr, E., Pintzinger, N. M., & Pfabigan, D. M. (2013). Happy and angry faces: Subclinical levels of anxiety are differentially related to attentional biases in men and women. *Journal of Research in Personality, 47*(4), 390–397.

Williams, L. M., Barton, M. J., Kemp, A. H., Liddell, B. J., Peduto, A., Gordon, E., & Bryant, R. A. (2005). Distinct amygdala-autonomic arousal profiles in response to fear signals in healthy males and females. *NeuroImage, 28*(3), 618–625.

Wood, G. E., & Shors, T. J. (1998). Stress facilitates classical conditioning in males, but impairs classical conditioning in females through activational effects of ovarian hormones. *Proceedings of the National Academy of Sciences U S A, 95*, 4066–4071.

Zeidan, M. A., Igoe, S. A., Linnman, C., Vitalo, A., Levine, J. B., Klibanski, A., . . . Milad, M. R. (2011). Estradiol modulates medial prefrontal cortex and amygdala activity during fear extinction in women and female rats. *Biological Psychiatry, 70*, 920–927.

Zorawski, M., Blanding, N. Q., Kuhn, C. M., & LaBar, K. S. (2006). Effects of stress and sex on acquisition and consolidation of human fear conditioning. *Learning and Memory, 13*, 441–450.

Zorawski, M., Cook, C. A., Kuhn, C. M., & LaBar, K. S. (2005). Sex, stress, and fear: Individual differences in conditioned learning. *Cognitive, Affective and Behavioral Neuroscience, 5*(2), 191–201.

Chapter 6
Sex Differences in Sensitivity to, and Salience of, Fear-Provoking Stimuli

Hitherto this point, discussion has focussed on components of the fear system that *respond* to the presence of threat as opposed to *how* we detect and interpret fear inducing stimuli. In this chapter, we examine the importance of two distinct but closely aligned concepts integral to the detection of fearful stimuli: Sensitivity (awareness of the presence of threat) and Salience (interpretation of the danger the threat poses). Intuitively, sex differences favouring heightened activity in women within these two areas, as with previous components, would be beneficial to their long-term fitness. These will each be considered in turn. Be advised that sensitivity and salience are examined via a wide array of different methods and readers unfamiliar with these may wish to consult Appendix A for task specific details.

6.1 Sensitivity

If women were more alert then men to cues of danger in the environment, this greater sensitivity in the form of a lower threshold for the activation of the fear system would provide a significant benefit in terms of additional time to select and execute avoidant action(s) which would keep them (and subsequently, their offspring) alive. It would, however, also result in the identification of a wider range of fear-eliciting stimuli and more frequent fear episodes (false positives). In extremis however, this could carry substantial costs in the form of hypervigilance, in which the environment is constantly scanned for threats, resulting in diminished cognitive capacity and increased distraction from essential ongoing survival tasks (Richards, Benson, Donnelly, & Hadwin, 2014). Indeed, women's vulnerability to PTSD and anxiety has been attributed to just such hypersensitivity to fear cues. Thus, while we might expect women to be more sensitive than men, the difference cannot be so great as to hinder survival functioning overall.

To assess system sensitivity, psychophysiological measures have been used in conjunction with the presentation of fear-provoking stimuli. Hence, the dependent

A. Campbell et al., *Sex Differences in Fear Response*, SpringerBriefs in Anthropology, https://doi.org/10.1007/978-3-030-65280-7_6

measure of sensitivity has been assessed in terms of autonomic response, rather than using psychophysical techniques to calibrate identification as a function of stimulus strength (although several studies have used tasks that include binary covert/overt presentation of fearful stimuli). An exception is a recent study measuring event related potentials (Forscher, Zheng, Zijung, Folstein, & Wen, 2016) which found that the linear discrimination of perceived fear intensity (operationalised as changes to perceived fear over a series of increasingly fearful morphed faces) occurred at 400–500 ms after a fearful stimuli was presented. Participants could also accurately detect changes in fearfulness (as low as a 6% change between two morphed faces). Unfortunately, sex differences in the speed and accuracy of these processes were not examined.

Breathing CO_2 enhanced air is a powerful way to induce extreme fear in order to measure psychophysiological responses. During normal breathing, CO_2 composes only 0.04% of inhaled air which is experimentally increased to as much as 35% CO_2 in some studies. Physiological responses to CO_2 inhalation are typically measured during the first few breaths. So far, no sex differences have been found for heart rate (Kelly, Forsyth, & Karekla, 2006; Leibold et al., 2013), changes in systolic or diastolic blood pressure (Leibold et al., 2013), skin conductance measures (Kelly et al., 2006; Nillni, Berenz, Rohan, & Zvolensky, 2012) or frontalis muscle tension (Kelly et al., 2006) in the presence of increasing concentrations of CO_2. Facial startle is also used to measure fear sensitivity. A sudden-onset auditory tone is delivered while viewing neutral images and electrodes above the orbicularis oculi (eye) muscle record the magnitude of facial startle. Most studies find no sex differences in baseline startle (Anokhin & Golosheykin, 2010; Hubbard et al., 2011; Kumari et al., 2008; Ludewig et al., 2003) although where they are found, they show a stronger response in women (Kofler, Muller, Reggiani, & Valls-Sole, 2001; Quevedo, Smith, Donzella, Schunk, & Gunnar, 2010).

The cognitive approach to sensitivity measurement employs a range of techniques assessing attentional bias for threat. These experimental tasks capitalise on the ‘pop out’ effect of feared over neutral stimuli and include the emotional Stroop task, dot probe, visual search (identification of a fear-inducing target among distractors), attentional blink (failure to detect a second fear-inducing target if it quickly follows the identification of a first), and eye tracking (eye movements are recorded in relation to fear-inducing stimuli). A meta-analysis of 4221 children’s (<18 years) performance on such tasks found a small ($d = 0.15$) to medium ($d = 0.54$) effect of anxiety on attentional bias toward threat but this was not moderated by sex (Dudeney, Sharpe, & Hunt, 2015).

The most commonly used attentional bias measure is the pure or modified dot probe task (Yiend, 2010). Two stimuli (one of which is threatening) appear simultaneously at different locations on a monitor (e.g. right and left). After a short exposure time both stimuli disappear and a probe appears at one of the two locations. Participants are asked to indicate as quickly as possible its spatial location. A bias toward threat is inferred when the reaction time on congruent trials (the probe appears at the same location as the threatening stimulus) is faster than incongruent trails (the probe appears at the location of the neutral stimulus). Typically, the bias index is computed by subtracting reaction times for congruent trials from incongruent trials so that a positive value indicates bias toward threat (attentional vigilance) and a negative value a bias away from threat (attentional avoidance). A meta-analysis of 2263 anxious and 1768 non-anxious individuals in studies using the dot probe and emotional Stroop tasks (Bar-Haim, Lamy, Pergamin, Bakermans-Kranenburg, & van IJzendoorn, 2007) found no attentional bias in non-clinical and low-vulnerability samples ($d = -0.01$) but a moderate bias in anxious samples ($d = 0.45$). Sex differences were not examined but given that women are generally more prone to anxiety than men (see Chap. 3), attentional bias measures might be expected to reveal greater threat vigilance in women.

Surprisingly few studies using the pure or modified dot probe task have compared men and women in the same experiment. Two studies found attentional vigilance only in high-anxious women (Tan, Ma, Gao, Wu, & Fang, 2011; Tran, Lamplmayr, Pintzinger, & Pfabigan, 2013). Even more specifically, Tan et al. found that the effect was restricted to highly anxious women only in the unconscious threat condition, with fear-stimulus presentation masked by binocular suppression (where high contrast visual distractions are presented to the dominant eye in order to suppress a meaningful image presented to the non-dominant eye). In non-anxious individuals of both sexes however (Pintzinger, Pfabigan, Tran, Kryspin-Exner, & Lamm, 2016; Tran et al., 2013) and in women (Carr, Scully, Webb, & Felmingham, 2016), threat avoidance bias is more common than vigilance. The evidence from cognitive approaches does not appear to support greater female sensitivity to threat cues in non-clinical populations.

In summary, although we may intuitively expect increased *sensitivity* to threat (offering more time and opportunity to consider and select evasive action) to be of adaptive benefit to women, threshold recalibrating carries costs which likely outweigh the advantages gained in terms of increased processing speed and safety. These costs include the false positive (Type 1) error of mobilising physiological, psychological and behavioural fear responses when they are not needed. A too-sensitive threat detection system would 'over-detect' threats resulting in an excessive number of false alarms and placing chronic energetic strain on the system. In the face of repeated and chronic stress, the balance between the interconnected Hypothalamic Pituitary Axis and the Hypothalamic Pituitary Gland reproductive axis can be disturbed, resulting in depleted estrogen levels and reduced ovarian function (Toufexis, Rivarola, Lara, & Viau, 2014), potentially harming long term reproductive fitness in women. These threats to successful female reproduction would outweigh benefits of frequent false positives and be expected to constrain the extent to which the fear system could be made more sensitive.

6.2 Salience

In psychology 'salience' is used to describe the identification of a stimulus that stands out from others because of its relevance for behaviour (Uddin, 2015), flagging it for priority processing in order to facilitate more rapid response behaviours. In the context of this discussion, it refers to how the level of threat a given stimulus poses (relative to other stimuli) is evaluated and prioritised for attention. As with sensitivity, an increased ability to identify threats would provide vital time for women to react by removing themselves (and their offspring) from harm's way. The salience system however also carries the same cost-benefit problem as highlighted previously in relation to sensitivity, in that while failure to detect a genuine threat could result in disastrous fitness consequences, a system that registers too many false positive may carry a heavy fitness burden in its own right. If the fear system is calibrated in such a way to impact women more than men, we may thus intuitively expect at least a small sex difference in this area.

A neural 'salience network' (overlapping with the fear system we identify here) has been identified jointly composed of the Anterior Insula and the Anterior Cingulate Cortex with connections to the amygdala, hypothalamus and periaqueductal gray. The Anterior Insula which coordinates interoceptive information about the internal visceral state is capable of detecting salience embedded in novel stimuli. This information is passed to the Anterior Cingulate Cortex which acts as an effector organ, initiating motor responses (Medford & Critchley, 2010). Frequently co-activated, these two structures appear to be responsible for switching attention between the default and the executive networks when a salient event is detected (Menon & Uddin, 2010). Both structures receive projections from the amygdala where the basolateral nucleus contains neurons which respond to salience as well as valence (Janak & Tye, 2015). Some of the gross effects of recalibrating salience would be similar to those of altering sensitivity. Speedier salience detection would direct attention faster to the source of danger, enabling a more finely tuned identification of the threat, better-tailored evasive action, and shorter response times. As with sensitivity, a too-liberal threshold for salience detection would increase the number of threats identified, possibly overwhelming the fear processing system and, in the long term, bringing about the consequences of chronic stress. The distinction between the constructs of salience and sensitivity requires further research but it can be suggested that while threat sensitivity refers to a bottom-up process by which a presented stimulus exceeds some form of identification threshold and is represented as threatening, salience implies a more cognitively active selection and prioritisation of cues to threat. This difference is subtle but nonetheless important.

The small quantity of available empirical evidence does not support a sex difference in this salience network. Women did not show more frequent joint activation of the Anterior Insula and Anterior Cingulate Cortex regions. Andreano, Dickerson, and Barrett (2014) found that women demonstrated smaller differences between novel and familiar negative images then men in both the Anterior Insula and subgenual Anterior Insular Cortex, suggesting a smaller difference in salience despite the differential novelty of the stimuli. Indeed, when viewing affective images, Moriguchi, Touroutoglou, Dickerson, and Barrett (2014) found greater connectivity between Anterior Insula and Anterior Insular Cortex in men. However, salience detection can occur at earlier points in the fear circuitry. The basolateral nucleus of the amygdala contains neurons whose firing in response to stimulus salience is correlated with autonomic activity and implicated in attention control (see Janak & Tye, 2015). Women, as reviewed earlier, show a stronger and more protracted amygdala response to threat. However, to the extent to which there is overlap between processes and structures involved in stimulus sensitivity (reflected in attentional bias studies) and salience, the absence of sex differences in the former, reviewed earlier, does not support differences between men and women.

'*Salience'* can however be interpreted in a less mechanistic way to refer to the *relevance* of a given stimulus for action. For example, making decisions between different courses of action involves the identification and weighting of salient variables. Risky behaviour can be viewed as the outcome of a decision-making process (not necessarily conscious) that assesses the benefits and costs of various outcomes multiplied by their relative probabilities of occurrence to arrive at a net value favouring one option. Once dominated by rational choice models, the field of decision making has increasingly incorporated the role of affect, especially in the domain of *risky decision making.* Fear can affect decision making by increasing uncertainty and lowering one's sense of personal control. Because fear increases the perception of risk, fear can also affect the costs, benefits and probabilities projected on alternative outcomes (Lerner, Li, Valdesolo, & Kassam, 2015). A stronger fear response in women would be expected to impact upon decision making by alerting them to the potentially negative consequences of risky decisions. Sex differences would not be expected on the weight assigned to positive outcomes and weaker effects would be expected for negative outcomes that did not involve physical injury (e.g. hypothetical financial loss). Thus, in line with our discussion so far, stronger sex differences would be expected where negative outcomes entailed increasing somatic risks.

Studies of risky decision making have concluded that women are more risk averse than men, although there is debate as to the magnitude of this effect, with the lowest values ranging from $d = 0.07$ to $d = 0.13$ (Byrnes, Miller, & Schafer, 1999; Croson & Gneezy, 2009; Nelson, 2015). These small effect sizes are derived from studies using cognitive decisions that are affectively 'cold' because they involve hypothetical scenarios that carry no personal consequences for the respondent (often referred to as System 2 thinking; Evans, 2003) and as such represent some degree of slower, deliberate and rational decision making. The sex difference is more marked when real-world behaviour such as investment choices for personal and pension assets are examined (Croson & Gneezy, 2009), and when the risks are more personally consequential, making them affectively 'hot' (Byrnes et al., 1999; Wake, Wormwood, & Satpute, 2020).

It has been argued that, in risky decision making especially, gut reactions often guide behaviour (Slovic, Finucane, Peters, & MacGregor, 2002) and decisions have a strong affective component (Loewenstein, Weber, Hsee, & Welch, 2001). Indeed,

Byrnes et al. (1999, p. 378) suggested that sex differences in fear "may explain gender differences in risk taking more adequately than the cognitive processes involved in the reflective evaluation of options". Lerner and Keltner (2000, 2001) suggest that fear increases pessimism in evaluating risky decisions, promoting risk-aversive choices. Following the World Trade Center attacks in 2001, women reported more fear, less anger and a higher estimation of future risk than men (Lerner, Gonzalez, Small, & Fischhoff, 2003). These sex differences in fear and anger explained between 60% and 80% of the sex difference in pessimistic risk estimation (where males were, as expected, less pessimistic than females).

There is persuasive evidence that women are more alert to the negative consequences of risky decisions than men. Women reported that they were less likely than men to take risks in the areas of gambling, recreation and health (although not in the less potentially injurious social domain). These sex differences were significantly mediated by women's ratings of the likelihood and severity of negative consequences (Harris, Jenkins, & Glaser, 2006). Even among skiers and snowboarders who regularly engage in risky activities, it is women's low sensitivity to punishment that predicts their risk-taking (Maher, Thomson, & Carlson, 2015). In women punishment sensitivity was strongly negatively predictive of risk taking ($r = -0.64$) while reward sensitivity positively predicted risk taking for men but not women. Meta-analysis shows a significant sex difference in punishment sensitivity favouring women, $d = -0.33$ over 17 studies, although its companion scale measuring sensitivity to rewarding or positive outcomes shows no sex difference, $d = 0.01$ (Cross, Copping, & Campbell, 2011). We may thus infer that fear of negative consequences promotes risk aversion in women.

Sex differences on the Iowa Gambling Task (IGT) also suggest that women are more loss-sensitive than men (van den Bos, Homberg, & de Visser, 2013). In this task, participants select a series of cards from one of four decks which vary in the size of rewards, penalties and the frequency of incurring them. Cards in decks A and B (disadvantageous decks) return rewards of $100 but occasionally (1/10) carry a large penalty resulting in a net loss of $250 over ten cards. Cards in decks C and D (advantageous decks) give smaller rewards ($50) but smaller losses resulting in an overall gain of $250 over ten cards. Although there is no sex difference in the ability to learn the rules that govern card values, women require on average an additional 40–60 trials before they equal the performance of men by consistently choosing from advantageous decks. The sex difference appears to result from women's greater sensitivity to (even small) losses in the context of the advantageous decks which causes them to shift to the disadvantageous decks. Similar effects have been found for the Risky Gains Task (Lee, Chan, Leung, Fox, & Gao, 2008) and the Cambridge Gambling Task (van den Bos et al., 2013). Suppressing this tendency to 'over-react' to loss is critical to implementing the better long-term strategy. While playing the IGT, women showed greater activation of the medial orbitofrontal cortex associated with the evaluation of immediate consequences, while men showed greater lateral Orbitofrontal Cortex activity associated with monitoring long-term rewards (reviewed in van den Bos et al., 2013).

Greater female sensitivity to negative outcomes is found on decision-making tasks in which participants chose between gambling (50% probability of winning $200 and 50% probability of winning nothing) and receiving a definite sum of money (ranging from $50–$120) (Powell & Ansic, 1997). In line with their greater risk aversion, women preferred the certain option. However, when the task was inverted so that it was framed in terms of potential losses, women took the 'risky' (gambling) option more than men. This outcome is congruent with women being more loss sensitive: They avoided the potential loss in the first version of the task (by taking the 'safe' option) and avoided the certain loss in the second version (by taking the 'risky' option).

The Balloon Analogue Risk Task (BART) is another widely used task for examining risky decision making. Participants pump up a computer-simulated balloon and each pump increases their prize points. If the balloon bursts, they lose that round's accumulated points, but participants can choose to 'bank' (protect) their point winnings at any time and move on to the next round. As one would expect, men typically pursue riskier strategies and earn more points on average than women ($d = 0.30$, Cross et al., 2011). The task has also identified several factors that enhance the sex difference in risky decision making. Stress reduced women's risk-taking but enhanced men's (Lighthall, Mather, & Gorlick, 2009). When instructions were framed to make punishing outcomes more relevant by emphasising that money could be lost if the balloons burst, women but not men showed a reduction in risk-taking (Gabriel & Williamson, 2010). A neuroimaging study of prefrontal cortex (Cazzell, Li, Lin, Patel, & Liu, 2012) revealed a sex difference following BART losses but not wins. Following losses, women showed significantly more dorsolateral prefrontal cortex activation bilaterally, indicative of conscious attempts to modulate emotional reaction. For both win and loss trials, women showed negative correlations between number of pumps (risk-taking) and dorsolateral prefrontal cortex activation suggesting that women's risky decisions were associated with a relaxation of executive control (dorsolateral prefrontal cortex activation). These correlations were positive for men. Relevant to the hypothesis that women's greater fear ultimately results from their mothering role, a BART study examined changes in risk-taking when participants were paired with various partners (Fischer & Hills, 2012). Men's risk taking increased only when they were paired with another man (suggesting competition), whereas women (but not men) were especially risk-averse when an image of a baby was displayed on the screen.

In summary, decision-making studies reveal that women are more attuned to the likelihood and severity of punishing outcomes than men. Increasing the salience of risk increases their risk aversion. Women's greater risk aversion is found in personally relevant behavioural decisions, as well as in laboratory paradigms including IGT, gambling choices and BART. In interpreting this effect, it is crucial to recognise that what is salient in these decisions is the *memory or projection* of previous experiences of loss, punishment or negativity and the valence strength that is attached to them. In scanning studies, activation of brain regions indicating salience are examined in response to the presence or absence of perceptual threat cues in visual images, as the individual is (presumptively) experiencing fear. Here there is little evidence of a sex difference. However, during decision-making, the individual calls up memories and/or projects constructions of fear in order to assign value to them and so inform decision making. This suggests that some sex differences in risk-averse decision making may derive from sex differences in memory for punishing events and thus linked with components of the fear system discussed earlier such as *subjective experience* and *time course* (see Chaps. 3 and 5).

To conclude our discussion in this chapter, it seems that sex differences in sensitivity to threatening stimuli, despite being intuitively expected, do not emerge. The potential fitness consequences of chronic stress induced by hypervigilance may be too costly for women to maintain. However, it should be noted that, of all the components of the fear system discussed so far, sensitivity appears to have the least available empirical evidence to synthesise conclusions from, and future work may lead to different conclusions. The same conclusion can be drawn to some degree regarding a traditional conceptualisation of 'salience' (i.e. the evaluation of an identified threat). However, if we consider salience to merely refer to the relevance of a stimulus to a future action/choice (i.e. how punishing this may be), then we are able to infer a consistent sex difference, with women generally being more pessimistic in the face of risky choices and erring toward more risk aversive strategies. As such, we can conclude that a salience component of the fear system would confer adaptive benefits to women (and their offspring) by promoting less risky decision making.

References

Andreano, J. M., Dickerson, B. C., & Barrett, L. F. (2014). Sex differences in the persistence of the amygdala response to negative material. *Social Cognitive and Affective Neuroscience, 9*, 1388–1394.

Anokhin, A. P., & Golosheykin, S. (2010). Startle modulation by affective faces. *Biological Psychology, 83*, 37–40.

Bar-Haim, Y., Lamy, D., Pergamin, L., Bakermans-Kranenburg, M. J., & van IJzendoorn, M. H. (2007). Threat-related attentional bias in anxious and nonanxious individuals: A meta-analytic study. *Psychological Bulletin, 133*, 1–24.

Byrnes, J. P., Miller, D. C., & Schafer, W. D. (1999). Gender differences in risk taking: A meta-analysis. *Psychological Bulletin, 125*, 367–383.

Carr, A. R., Scully, A., Webb, M., & Felmingham, K. L. (2016). Gender differences in salivary alpha-amylase and attentional bias towards negative facial expressions following acute stress induction. *Cognition and Emotion, 30*(2), 315–324.

Cazzell, M., Li, L., Lin, Z.-J., Patel, S. J., & Liu, H. (2012). Comparison of neural correlates of risk decision making between genders: An exploratory fNIRS study of the Balloon Analogue Risk Task (BART). *NeuroImage, 62*, 1896–1911.

Croson, R., & Gneezy, U. (2009). Gender differences in preferences. *Journal of Economic Literature, 47*(2), *448–474*.

Cross, C. P., Copping, L. T., & Campbell, A. (2011). Sex differences in impulsivity: A meta-analysis. *Psychological Bulletin, 137*(1), 97–130.

Dudeney, J., Sharpe, L., & Hunt, C. (2015). Attentional bias towards threatening stimuli in children with anxiety: A meta-analysis. *Clinical Psychology Review, 40*, 66–75.

Evans, J. S. B. T. (2003). In two minds: Dual-process accounts of reasoning. *Trends in Cognitive Sciences, 7*(10), 454–459.

Fischer, D., & Hills, T. T. (2012). The baby effect and young male syndrome: Social influences on cooperative risk-taking in women and men. *Evolution and Human Behavior, 33*, 530–536.

Forscher, E. C., Zheng, Y., Zijung, K., Folstein, J., & Wen, L. (2016). Decomposing fear perception: A combination of psychophysics and neurometric modelling of fear perception. *Neuropsychologia, 91*, 254–261.

Gabriel, K. A., & Williamson, A. (2010). Framing alters risk-taking behaviour on a modified Balloon Analogue Risk Task (BART) in a sex-specific manner. *Psychological Reports, 107*(3), 699–712.

Harris, C. R., Jenkins, M., & Glaser, D. (2006). Gender differences in risk assessment: Why do women take fewer risks than men? *Judgment and Decision making, 1*(1), 48–63.

Hubbard, C. S., Ornitz, E., Gaspar, J. X., Smith, S., Amin, J., Labus, J. S., ... Naliboff, B. D. (2011). Modulation of nociceptive and acoustic startle responses to an unpredictable threat in men and women. *Pain, 152*, 1632–1640.

Janak, P. H., & Tye, K. M. (2015). From circuits to behaviour in the amygdala. *Nature, 517*, 284–292.

Kelly, M. M., Forsyth, J. P., & Karekla, M. (2006). Sex differences in response to a panicogenic challenge procedure: An experimental evaluation of panic vulnerability in a non-clinical sample. *Behaviour Research and Therapy, 44*, 1421–1430.

Kofler, M., Muller, J., Reggiani, L., & Valls-Sole, J. (2001). Influence of gender on auditory startle responses. *Brain Research, 921*, 206–210.

Kumari, V., Aasen, I., Papadopoulos, A., Bojang, F., Poon, L., Halari, R., & Cleare, A. J. (2008). A comparison of prepulse inhibition in pre- and postmenopausal women and age-matched men. *Neuropsychopharmacology, 33*, 2610–2618.

Lee, T. M., Chan, C. C., Leung, A. W., Fox, P. T., & Gao, J. H. (2008). Sex-related differences in neural activity during risk taking: An fMRI study. *Cerebral Cortex, 19*(6), 1303–1312.

Leibold, N. K., Viechtbauer, W., Goossens, L., De Cort, K., Griez, E. J., Myin-Germeys, I., ... Schruers, K. R. J. (2013). Carbon dioxide inhalation as a human experimental model of panic: The relationship between emotions and cardiovascular physiology. *Biological Psychology, 94*, 331–340.

Lerner, J. S., Gonzalez, R. M., Small, D. A., & Fischhoff, B. (2003). Effects of fear and anger on perceived risk of terrorism: A national field experiment. *Psychological Science, 14*(2), 144–150.

Lerner, J. S., & Keltner, D. (2000). Beyond valence: Toward a model of emotion-specific influences on judgment and choice. *Cognition and Emotion, 14*, 473–493.

Lerner, J. S., & Keltner, D. (2001). Fear, anger, and risk. *Journal of Personality and Social Psychology, 81*, 146–159.

Lerner, J. S., Li, Y., Valdesolo, P., & Kassam, K. S. (2015). Emotion and decision making. *Annual Review of Psychology, 66*, 799–823.

Lighthall, N. R., Mather, M., & Gorlick, M. A. (2009). Acute stress increases sex differences in risk seeking in the Balloon Analogue Risk Task. *PLoS One, 4*, e6002.

Loewenstein, G. F., Weber, E. U., Hsee, C. K., & Welch, N. (2001). Risk as feelings. *Psychological Bulletin, 127*(2), 267–286.

Ludewig, K., Ludewig, S., Seitz, A., Obrist, M., Geyer, M. A., & Vollenweider, F. X. (2003). The acoustic startle reflex and its modulation: Effects of age and gender in humans. *Biological Psychology, 63*, 311–323.

Maher, A. M., Thomson, C. J., & Carlson, S. R. (2015). Risk-taking and impulsive personality traits in proficient downhill sports enthusiasts. *Personality and Individual Differences, 79*, 20–24.

Medford, N., & Critchley, H. D. (2010). Conjoint activity of anterior insular and anterior cingulate cortex: Awareness and response. *Brain Structure and Function, 214*, 535–549.

Menon, V., & Uddin, L. Q. (2010). Saliency, switching, attention and control: A network model of insula function. *Brain Structure and Function, 214*(5-6), 655–667.

Moriguchi, Y., Touroutoglou, A., Dickerson, B. C., & Barrett, L. F. (2014). Sex differences in the neural correlates of affective experience. *Social Cognitive and Affective Neuroscience, 9*, 591–600.

Nelson, J. (2015). Are women really more risk-averse than men? A re-analysis of the literature using expanded methods. *Journal of Economic Surveys, 29*(3), 566–585.

Nillni, Y. I., Berenz, E. C., Rohan, K. J., & Zvolensky, M. J. (2012). Sex differences in panic-relevant responding to a 10% carbon dioxide-enriched air biological challenge. *Journal of Anxiety Disorders, 26*, 165–172.

Pintzinger, N. M., Pfabigan, D. M., Tran, U. S., Kryspin-Exner, I., & Lamm, C. (2016). Attentional biases in healthy adults: Exploring the impact of temperament and gender. *Journal of Behavior Therapy and Experimental Psychiatry, 52*, 29–37.

Powell, M., & Ansic, D. (1997). Gender differences in risk behaviour in financial decision-making: An experimental analysis. *Journal of Economic Psychology, 18*(6), 605–628.

Quevedo, K., Smith, T., Donzella, B., Schunk, E., & Gunnar, M. (2010). The startle response: Developmental effects and a paradigm for children and adults. *Developmental Psychobiology, 52*, 78–89.

Richards, H. J., Benson, V., Donnelly, N., & Hadwin, J. A. (2014). Exploring the function of selective attention and hypervigilance for threat in anxiety. *Clinical Psychology Review, 34*, 1–13.

Slovic, P., Finucane, M. L., Peters, E., & MacGregor, D. G. (2002). The affect heuristic. In T. Gilovich, D. Griffin, & D. Kahneman (Eds.), *Heuristics and biases: The psychology of intuitive judgment* (pp. 397–420). Cambridge and New York: Cambridge University Press.

Tan, J., Ma, Z., Gao, X., Wu, Y., & Fang, F. (2011). Gender difference of unconscious attentional bias in high trait anxiety individuals. *PLoS One, 6*(5), e20305.

Toufexis, D., Rivarola, M. A., Lara, H., & Viau, V. (2014). Stress and the reproductive axis. *Journal of Neuroendocrinology, 26*(9), 573–586.

Tran, U. S., Lamplmayr, E., Pintzinger, N. M., & Pfabigan, D. M. (2013). Happy and angry faces: Subclinical levels of anxiety are differentially related to attentional biases in men and women. *Journal of Research in Personality, 47*(4), 390–397.

Uddin, L. Q. (2015). Salience processing and insular cortical function and dysfunction. *Nature Reviews Neuroscience, 16*, 55–61.

van den Bos, R., Homberg, J., & de Visser, L. (2013). A critical review of sex differences in decision-making tasks: Focus on the Iowa Gambling Task. *Behavioural Brain Research, 238*, 95–108.

Wake, S., Wormwood, J., & Satpute, A. B. (2020). The influence of fear on risk taking: A meta-analysis. *Cognition and Emotion, 34*(6), 1143–1159. https://doi.org/10.1080/02699931.2020.1731428.

Yiend, J. (2010). The effects of emotion on attention: A review of attentional processing of emotional information. *Cognition and Emotion, 24*, 3–47.

Chapter 7
Discussion and Conclusions

7.1 Summarising the System

As we approach the final discussion of the fear system, it is best to begin with a functional account of how these mechanisms work together as an integrated system. Inspection of sex differences across the components of the fear system discussed in the previous chapters (2–6) has provided differential support for their candidacy as targets of selection in enhancing women's reproductive success. Studies of risky decision-making (including laboratory paradigms as well as personal decisions) reveal that women are more risk averse, assigning a greater weight to potential costs compared to men and modifying their behaviour more than men in response to punishment (but not reward; Cross, Copping, & Campbell, 2011). Women also report *subjectively experiencing* fear more frequently and more intensely than men do. This greater self-reported fear in women is likely to be underpinned by two connected components of the fear system: *strength* and *time course*. Neuropsychological studies have provided persuasive evidence for a stronger and better integrated response across components of the fear system in women especially in response to angry faces and direct threat. When danger is encountered, this stronger aversive neural response will strongly motivate attempts to avoid or escape. The more extended *time course* of fear enhances women's ability to anticipate danger and take precautionary measures. Women's slower extinction of acquired fears means that learned fears are more resistant to replacement and this conservative bias errs on the side of mother and offspring survival. Greater sustained fear in women (the protracted watchfulness associated with detecting possible danger) together with evidence from potentiated startle studies (which effectively 'cue' and thereby extend phasic fear) afford a longer duration to engage frontal regions in avoidance planning, rather than a reliance on periaqueductal grey control mediated automatic fight-and-flight responses.

The fear system is well conserved across species and conscious, *subjective awareness* of emotion is not necessary for the fear circuit to operate successfully

A. Campbell et al., *Sex Differences in Fear Response*, SpringerBriefs in Anthropology, https://doi.org/10.1007/978-3-030-65280-7_7

in avoiding danger: 'feelings' of fear may be a *product* of the fear system, but facilitating our survival is its primary *function* (LeDoux, 2012). The possibility that women are more aware of fear at the time that they experience it is suggested by their reports of greater intensity and frequency of subjective fear, but the limits of self-report measures are important here. Answering the question 'How intensely or frequently do you experience fear?' implicitly requires the respondent to recollect previous episodes of fear and produce an average value: it is essentially a memory task, and women's spontaneous recall of autobiographical memory appears to be more negative in content than men's (Young, Bellgowan, Bodurka, & Drevets, 2013). Furthermore, amygdala activation is correlated with strength of encoding of emotional stimuli (Canli, Desmond, Zhao, & Gabrieli, 2002). Women's greater self-report of fear might therefore reflect increased 'accessibility of fearful events' in memory. There does not appear to be a sex difference in anterior insulae activity associated with emotional and visceral interoception, which suggests that the 'on-line' emotional tracking of fear is distinct from the information required for self-reported fear estimates.

There is little evidence from psychophysiological or attentional bias studies that the sexes differ in fear *sensitivity* or threshold for the identification of threat. Such a recalibration would risk the elaborated processing of non-critical fears that could result in hypervigilance and increased stress which have been tied to reproductive failure in women. The allied concept of salience focuses on the identification of stimuli that are flagged for priority processing because of their heightened implications for action. Neuropsychological studies do not support stronger activity in women's 'salience network' composed of the anterior insulae and the anterior cingulate cortex.

The components of the fear system which are most likely the products of selection then, are *strength* (of the neural response to threat and the modification of behaviour) and *time course* (persistence of this modification). *Salience* and *sensitivity* are the least likely candidates. Evidence for sex differences in the *subjective experience* of fear is more ambiguous to interpret although does seem to at least suggest that women experience it more strongly than men.

7.2 Women as Survivors

The crux of the 'staying alive' hypothesis is that a stronger fear system increases the survivability of women and thus their overall fitness through continued offspring provisioning and protection (see Chap. 1). Women do generally stay alive for longer than men. Evidence from population studies on mortality rates suggest that women have an advantage in overall life expectancy across time and geographical location (Austad, 2006; Lemaire, 2013) as well environment type, socioeconomic status and whether or not the cause of the mortality was internal or external (Kruger & Nesse, 2006). This difference also exists even in the youngest in society, with child mortality rates at ages 0–5 years also showing a survival advantage for girls

(Drevenstedt, Crimmins, Vasunilashorn, & Finch, 2008; United Nations, 2011). The sex difference in mortality is also evident in most mammalian species (Clutton-Brock & Isvaran, 2007) in which male-male competition is normative. The suggested reasons for the sex differences in mortality are numerous and mostly biological in nature. Studies on the immune system suggest that estrogens have several health promoting effects while testosterone can act as an immunosuppressant (Haring et al., 2010; Waldron, 1983; Xing, Nozell, Chen, Hage, & Oparil, 2009). There are also suggested disease resistance factors associated with having two X chromosomes (Morris & Harrison, 2009). While readers may point out that male mortality is likely to be higher due to reasons depicted in Chap. 2 (i.e. extrinsic risk taking), it should be noted that in studies examining populations where the behaviours of men and women are very similar in terms of low risk and risk taking (for instance, in cloistered religious communities) women still seem to live longer than men (Brønnum-Hansen & Juel, 2001; Lindahl-Jacobsen et al., 2013; Luy, 2003).

There is evidence to suggest that women generally are also better at surviving across many different extreme environmental conditions, not just general and everyday risky situations. Zarulli and colleagues (Zarulli et al., 2018) argue that even in the most extreme conditions (famines, epidemics and slavery) the lower mortality rates and higher life expectancies of women remain almost ubiquitous. These extreme situations represent both internal and external sources of mortality, but in either case, women appear more robust to their impact. This begs the question of whether selection over our evolutionary past has favoured mechanisms that increase female survivorship (and would support the biological arguments mentioned above). If this was the case, an enhanced fear system may be one such mechanism that contributes to increased survivorship in women. In times of extreme crisis, a system that continues to inhibit risk taking behaviour may make the difference between survival and death for women. This could take the form of being increasingly frugal with resources in times of shortfall and being more conscious and avoidant of health risks in a time of increased pathogenesis. Such a mechanism may have made all the difference in contributing to keeping mothers and their children alive through times of adversity. An enhanced fear system would still be adaptive even in the most adverse conditions.

7.3 Staying Alive and Competing Perspectives

The essence of the 'staying alive' account offered within this thesis of women's heightened fear response (Campbell, 1999, 2013) stands in marked contrast to another evolutionary proposal, the 'tend-and-befriend' hypothesis (Taylor et al., 2000). Both theories emphasise the key importance of the mother in ensuring infant survival, but they differ in women's proposed response to danger. Although Taylor et al. (2000, p. 415) acknowledge that "In both sexes, threat triggers sympathetic-adrenal-medullary (SAM) and hypothalamic-pituitary-adrenal (HPA) activation...", they argue that the sympathetic nervous system fight-or-flight response that is typical of men has been down-regulated in women. The anxiety-reducing action of the peptide oxytocin is invoked to explain the suppression of women's physiological stress response, their emotional disengagement, and their tendency to seek mutual social support from other women ('befriend') and to nurture ('tend') their young during threat. The survival advantages of this passive strategy presumably lie in numbers, specifically that other women will act as distractors to predators or conspecifics resulting in the survival of one's own offspring.

The tend-and-befriend hypothesis is also challenged by an empirical problem. Oxytocin has been widely hailed for its anxiety-reducing properties and early research confirmed a reduction in amygdala activation to fearful faces and threat-related scenes (see Lischke et al., 2012). Although this research was performed on men, oxytocin's up-regulation by estrogen and its involvement in parturition, lactation and mothering suggested a special relevance to females. However, two studies (Domes et al., 2010; Lischke et al., 2012) have now demonstrated that in women, administration of intranasal oxytocin significantly *increases* rather than decreases amygdala activation in response to fear-inducing faces and scenes. Oxytocin also *increases*, rather than decreases, the startle response to sustained fear, again demonstrating anxiogenic effects (Grillon et al., 2013).

Reversing the presumed causality of the tend-and-befriend hypothesis, the more egalitarian and emotionally bonded friendships between women that are attributed to oxytocin may actually be an indirect *result* of women's enhanced fear which steers them away from injurious forms of intrasexual competition. For example, among Gombe chimpanzees over a 44-year period the average number of agonistic interactions among females was 46.1, compared to 249.2 for males (Foerster et al., 2016). There was not a single successful rank challenge by females who, unlike males, queued rather than fought for status. In human women, intrasexual competition for status has a destructive effect on female relationships (Lee, Kesebir, & Pillutla, 2016). Female cliques rarely show the clear hierarchical structure typical of male groups. Young women want to 'fit in' rather than stand out, which explains the paradoxical finding that girls who are nominated as 'popular' (visible, charismatic) tend to be less well-liked as friends (Closson, 2009; Eder, 1985). This is not to argue that women do not compete but rather that their competition is typically non-confrontational and indirect (Campbell, 2013, 2015).

Contrary to the tend-and-befriend hypothesis, the 'staying alive' hypothesis proposes that heightened, rather than diminished, attention to threat has been adaptive for women (Campbell, 1999, 2013), concordant with evidence of women's higher levels of fear reviewed in this article. Over 95% of new mothers (and 80% of fathers) experience repeated thoughts about potential harm to their infants over the first 3 months of life (Leckman et al., 1999), often deriving from concern about disease, accidental neglect, and 'stranger danger'. These pessimistic preoccupations motivate precautionary behaviours aimed at enhancing child survival (reviewed in Hahn-Holbrook et al., 2011). In emphasising the role of fear in prompting distal precautions (e.g. avoiding sharp drops, bodies of water, etc.), fear also pays a role in proximate maternal defence (mothers' willingness to actively aggress against threatening conspecifics). Animal studies identify maternal defence against infanticidal competitors as the 'other side of the coin' of mother–infant bonding. Consistent with the role of fear in guiding mothering, willingness to mount a risky attack in the service of saving an infant's life depends on temporary fear *reduction* (Campbell, 2008). While oxytocin inhibits aggression directed towards pups, it simultaneously enhances maternal attacks on intruders (Debiec, 2005; Pedersen, 2004) and is associated with low levels of fear (Gammie, Negron, Newman, & Rhodes, 2004; Hard & Hansen, 1985). In a unique human study using a laboratory aggression paradigm, breast-feeding mothers (associated with higher oxytocin levels) demonstrated greater aggression against a provocative confederate than did formula-feeding mothers or nulliparous women. While they were aggressing, the lactating women also showed lower systolic blood pressure and mediation analyses indicated that reduced arousal can disinhibit female aggression (Hahn-Holbrook, Holt-Lunstad, et al., 2011). In the presence of a hypothetical male representing 'stranger danger', parents evaluate them as more formidable and threatening than non-parents (Fessler, Holbrook, Pollack, & Hahn-Holbrook, 2014). Making parenting roles more salient increases parents' perception of risk and their preference for risk-averse choices while decreasing their trust in strangers (Eibach & Mock, 2011).

This volume has focused on evolutionary principles as the basis for understanding sex differences in fear. But sex differences have traditionally been explained in terms of cultural transmission of gender roles. Socialisation theories propose that boys and men are strongly encouraged to conform to the masculine role which discourages the manifest expression of fear. Biosocial theory (Wood & Eagly, 2012), one of the most influential of these approaches, initially rejected a role for evolution in the content of gender roles (Wood & Eagly, 2002) but now accepts that sex differences in minimum parental investment have implications for greater male competitiveness and dominance, and the suppression of the public display of vulnerable emotions. Familial and cultural processes can encourage men to deny or minimise the experience of fear, to redefine moderate levels of fear as pleasurable rather than aversive, or even to tolerate aversive levels of fear in the interests of 'masculine' self-presentation.

Regarding the first of these possibilities, one way to examine gendered minimisation of fear reports is to compare men's and women's ratings after observing their behaviour in the same situation. Several months after naturally-occurring disasters, women report more symptoms of PTSD. Although this may be the result of a women's greater vulnerability to stress-related disorders in general, there is also evidence that women experience these events as more frightening. A study of bank employees all of whom had been exposed to a robbery (Christiansen & Hansen, 2015) found that the largest sex differences were in the reported severity of peri-traumatic variables (fear, horror, helplessness, panic and dissociation). Furthermore, and as reviewed above, some experimental studies have measured and statistically controlled for gender role or have informed participants (incorrectly) that the veracity of their fear ratings can be objectively checked. Neither of these eliminated the significant effect of biological sex. Although gender conformity doubtless plays a role in self-presentation, evidence suggests that exaggeratedly low self-reports of fear by men cannot wholly explain the sex difference in fear. Men's reports are after all congruent with a large body of research demonstrating their greater behavioural risk-taking. It is also salutary to bear in mind that, despite the assumption of *male* under-reporting of fear, Arrindell and Buikhuisen (1992) found higher lie scores among women on the Fear Survey Schedule (as well as higher fear ratings). Self-reported fear of crime can also be subject to female exaggeration as well as male minimisation (Sutton, Robinson, & Farrall, 2011).

Another cultural explanation is that men may be socialised to interpret threat differently than women. Threat from a rival male or group may be interpreted as a challenge to which the appropriate response is hostility rather than fear. Men more than women view their aggression in instrumental terms as a means of controlling another person's behaviour, rather than as an expressive demonstration of their distress (Campbell, 1999; Campbell & Muncer, 1987). This is concordant with research suggesting a distinction between women's focus on internal feeling states and men's focus on external sensory data and their implications for action during negative emotional processing (Wager, Phan, Liberzon, & Taylor, 2003). Men respond to emotional images with activity in the primary visual cortex and fusiform gyrus indicative of an exterior focus during emotional arousal (Moriguchi, Touroutoglou, Dickerson, & Barrett, 2014) while women experience emotion as an internal somatic state. Sex differences in the interpretation of and response to threat may themselves have evolutionary roots. The male warrior hypothesis (McDonald, Navarette, & Van Vugt, 2012) proposes that males have played a greater part in intergroup aggression (due to the greater value of female survival) responding to threat with antagonism, while the reproductive costs to females in terms of outgroup sexual assault have supported a role for enhanced vigilance and avoidance.

Finally, it has been suggested that men's greater involvement in risky activities stems from the fact that masculinity, more than femininity, is viewed as a social achievement that requires public displays (Vandello & Bosson, 2013) exemplified for instance with 'Risk taking' being one of the eight subscales on the Conformity to Masculine Norms Inventory (Hsu & Iwamoto, 2014). Risk taking, a behaviour that is hard to successfully fake safely and thus an honest sexual signal, represents the most persuasive sign of manhood. Men who experience a weak sense of their own masculinity ('masculine discrepancy stress') whether as a trait or as an experimentally induced state, are especially likely to engage in compensatory risky behaviours. Despite their dangers, high-risk behaviours may be used as public demonstrations of masculinity among those with doubts about their gender status. For example, studies have shown that gender role stress can be associated with financial risk taking (Weaver, Vandello & Bosson, 2013) and high-risk sexual practices (Reidy, Brookmeyer, Gentile, Berke, & Zeichner, 2016). However, such studies often include only male participants and demonstrate correlation rather than cause. In addition, the ubiquity of men's greater risk taking across cultures that have differing criteria for masculinity (Costa, Terracciano, & McCrae, 2001; Hofstede, 1991; Wood & Eagly, 2002) further suggests that compensatory behaviour alone is unlikely to be a complete explanation of men's riskier behaviour. If men's fear was equal to that of women, the force of masculine role conformity would have to be extremely strong to motivate participation given the inherent danger of these activities. Many of these behaviours, whether pre-planned (e.g. extreme sports) or spontaneous (e.g. high-speed driving), are undertaken voluntarily and on occasions when they are not subject to public scrutiny.

There can be no doubt about the need to incorporate cultural transmission into our understanding of human evolution. Reciprocally, an understanding of any gendered/sex roles which are found to be stable across history, geography and culture requires consideration of their evolutionary genesis. Evolutionary psychologists now examine culture at the micro level in terms of the transmission of gender roles between individuals and the responsivity of these roles to changing ecological conditions (Cross, Cyrenne, & Brown, 2013). The study of evolved 'content biases' in human cognition (Boyd & Richerson, 1988; Mesoudi, 2016) addresses ways in which multidimensional and continuous variables (such as traits associated with gender roles) tend to be simplified into discrete and universal entities for the purposes of description and communication. The primate ability for conceptual and observational learning of gender roles is part of our evolutionary heritage that increases behavioural variance and adjustment to local conditions, rather than an alternative explanation of sex differences.

7.4 Limitations of the Current Literature and Future Directions

While this thesis presents compelling evidence for the adaptive advantages of heightened fear for women, there are several key limitations with existing research to address, as well as several areas that require additional research attention.

7.4.1 Methodological Issues

One of the major barriers to the study of the fear system is the availability of techniques which can provide the necessary evidence. Much of the work drawn upon in this volume includes neuroimaging studies: as discussed earlier, these pose several difficulties such as choosing appropriate stimuli to interpreting the relationship between structure and function. This is particularly problematic for studying fear, which earmarks a central role for the amygdala, its sub regions and connections to other cortical structures. Note that this thesis does not argue that the amygdala is 'for fear', merely that it has a critical role in fear. Nonetheless, it is noteworthy that research into sex differences in this region has lacked consistency in findings over the past 15 years (Andreano, Dickerson, & Barrett, 2014), a conclusion that can be extended to other important brain areas also. On the subject of neuroimaging studies, it is also important to highlight that many of these studies are relatively small in terms of their sample size (largely due to the time, cost and methodological limitations of such research) and are generally drawn from WEIRD populations (Western, Economic, Industrial, Rich and Democratic). While the results of these studies are promising and insightful, there are limits to their generalisability and thus conclusions regarding human universalities must be made with some degree of caution. Time will no doubt clarify the role of these cortical regions as scanning techniques become increasingly sophisticated, but for the time being, we must continue to be cautious with our treatment of findings when it comes to fear.

Similarly, there is also a prevailing issue around sampling in studies on fear conditioning and extinction. Several authors have noted that studies in these areas are likely to be samples of males only, and the literature has a surprising paucity of studies that directly compare males and females (Cover, Maeng, Lebron-Milad, & Milad, 2014; Merz, Kiner, & Wolf, 2018). Studies that do so, often find non-significant differences (see Chap. 5) or often don't report results by sex. This makes the literature in this area harder to interpret and there is a need to remedy this in future work.

7.4.2 Hormones and Fear

Readers may recall an earlier note regarding our limited attention to the potential effects of hormones on the fear system, with brief exceptions in Chaps. 4 and 5 where we (very briefly) examined the effects of estrogen, testosterone and oxytocin. We acknowledge that there is a broader literature around these hormones (and that of others i.e. progesterone, stress hormones such as cortisol etc.) and their wider role, particularly in the areas of fear extinction research. However, it is beyond the scope of this review to discuss all of them in detail. Interested readers who wish to know more regarding sex differences in these areas are invited to consult the works of Day and Stevenson (2019), Merz et al. (2018) and Verlasco, Florido, Milad, and Andero (2019) for more comprehensive overviews of this body of research.

What is worth noting from the literature reviewed (and a key finding in the reviews suggested above) is the consistent finding that estrogen has a role to play in fear conditioning and extinction. While this is an interesting finding, the current body of literature, particularly in studies where circulating hormone levels are not accounted for, makes the nature of this mechanism difficult to truly appreciate at this time. This is even more problematic when we consider that oral contraceptives (which keep estradiol and progesterone levels low throughout) are also not always factored in. Given that estradiol levels naturally fluctuate at different points during the menstrual cycle (increasing levels during the Follicular phase and lower during the Luteal phase) one could rightly question how this system would be evolutionarily adaptive. That is, why would it be of benefit to women to be more afraid during the days just prior to the onset of ovulation and less afraid for the days after it, given that is cycle repeats throughout the lifespan until menopause? How could it be adaptive to be less risk averse for only approximately half of the cycle? This is a question that requires some attention in future research. Speculatively, this may be linked to copulatory behaviours in women.

Copulation is an intimate act that requires us to overcome inhibition. From an evolutionary perspective, selection pressures act to inhibit sexual impulses in women to provide more time to assay their choice of sexual partner in terms of suitability (due to the imbalance in parental investment between men and women; Bjorklund & Kipp, 1996). It is not adaptive for women to take reproductive risks via poor mate choice. There is also a risk of physiological harm to women through acts of copulation (Franklin, 2010) which women seem attuned to. For example, Herold and Mewhinney (1993) showed that only 7% of men feared physical harm when alone with a stranger, contrasting to 52% of women. Sex thus carries potential risks for women that could have fitness consequences. However, if fear prevented copulation at critical reproductive junctures (such as ovulation) then a women's reproductive success would be adversely affected. To not reproduce would mean lineage extinction. It may thus be the case that the increase in estradiol just prior to the onset of ovulation (the periovulatory phase) and the days succeeding it (i.e. when conception is possible) reduces fear sufficiently to increase engagement in successful copulatory activity that would increase the probability of conception in this key window. To our knowledge, there is no direct empirical examination of this from the perspective of examining fear. However, evidence from research on sexual behaviour in women suggests that sexual activity is at its highest during the periovulatory phase of the menstrual cycle (Bullivant et al., 2004), particularly in women without a partner (Caruso et al., 2014). These heightened sexual activities may be partly as a consequence of an estradiol mediated fear system. Future work is required to fully explain the role of estradiol and fear from an evolutionary perspective, however.

7.4.3 Adaptive Variation

It will also have not escaped the notice of readers that little of this review is dedicated to research from the animal literature (with a few small exceptions). This is not to suggest that such literature is unimportant, only that the aim was to retain the research focus of this review on fear in humans, more specifically, what one might consider to be the 'modal' man and woman. Readers should be aware however that there is likely to be some element of plasticity in the fear system. Like many human traits and characteristics, there is likely to be a degree of adaptive variation linked to local environmental contexts. Indeed there is a large body of work from the literature on human life history strategies that suggest adaptive variation in a range of behaviours relevant to reproduction such as aggression, mating strategies, risk taking, sensation seeking (Copping & Campbell, 2015; Copping, Campbell, & Muncer, 2013; James, Ellis, Schlomer, & Garber, 2012) in response to levels of localised environmental harshness and unpredictability (sources and variation in mortality which naturally vary across human environments and populations—Ellis, Figueredo, Brumbach, & Schlomer, 2009) which would impact upon rates of inter and intrasexual competition. As reviewed earlier, these domains all show expected sex differences that may be underpinned by sex differences within the fear system. While adaptive variation may be at least partially responsible for creating individual differences in the levels of these behaviours, it would be expected that many pressures in the environment that increase or decrease levels of behaviour would act upon both sexes simultaneously. Indeed, the correlation between men and women's levels of a behaviour such as aggression/criminality is very strong (Campbell, Muncer, & Bibel, 2001) while maintaining the overall sex difference. While there may be variation in fear related behaviour, we would hypothesise that the sex difference in fear would be maintained. Similarly, we would also expect variation within the fear system to take place at different points over the reproductive lifespan and that levels of competitive engagement will vary depending on acute needs (Stockley & Campbell, 2013). To our knowledge, fear has yet to be examined specifically in a life history context but its inclusion in such studies would be a fruitful avenue for future research.

7.4.4 *Future Work*

This review ties together several disparate studies indicating the presence of sex differences in key identified areas of the fear system that support the theoretical proposition that such differences may have evolved as a functional survival adaptation. Readers will no doubt note the presence of literature gaps and a lack of explicit tests of some of these key ideas. Specific experimental work is lacking and thus while this synthesis shows tentative support for the staying alive hypothesis, more work is needed to make truly causal conclusions. This review should serve as a foundation from which to guide further work in the exploration of sex and fear. From the staying alive hypothesis, numerous, testable evolutionarily driven hypotheses can be generated. Are women behaviourally more cautious in response to the same stimuli under conditions of fear than men? Does aversive behaviour persist for longer in females than in males once the fear/threat has passed? Are there differential effects between women who do and do not already have offspring? Such questions could be addressed with experimental paradigms and we hope that future researchers will take up the mantle so that we might better understand the role of fear and its impact upon evolutionarily relevant behaviours.

7.5 Conclusions

We have argued that a more robust fear system enhances the survival of women and their children. Sex differences in reaction to threat are expected to feed forward to inform anticipatory and preventative behaviours. In a dangerous encounter, such as a narrow escape from a predator, women's stronger and more protracted systemic response forms the basis for a durable association between contextual cues and danger that is resistant to extinction. Because fear intensity enhances memory consolidation, both the experience of fear and the context of its occurrence are likely to be well remembered, informing decision-making about preferred routes or foraging sites. A return to this context (and by generalisation, other similar contexts) generates a low-level 'background' fear supporting vigilance and heightened reactivity to the appearance of danger. This watchful preparedness may include the mental construction of hypothetical futures that can inform avoidance and escape strategies.

The benefits of fear in steering women and their children away from physical injury may carry ancillary costs. While these negative consequences cannot theoretically outweigh the evolutionary benefits in the long term (a false positive may mean the loss of an immediate advantage but a false negative may result in injury or death), they are nonetheless unpleasant. Women are more than twice as likely as men to be diagnosed with an anxiety disorder (Bekker & van Mens-Verhulst, 2007) and twice as likely to suffer from panic attacks (de Jonge et al., 2016). PTSD is also twice as common in women with female sufferers being more likely than men to report acute peritraumatic responses including intense fear, helplessness, horror, intrusive thoughts, avoidance, panic, anxiety, and dissociation (Olff, Langeland, Draijer, & Gersons, 2007). Increasingly, biological sex is being recognised as a key factor in understanding and treating psychopathology (Del Giudice & Ellis, 2016).

In summary, although women and men are equally sensitive to danger cues in the environment, once threat is detected, women show a stronger and more protracted fear response. This may explain why women engage less often in risky activities, report more intense fear, assign a greater weight to the costs associated with risky decisions, are slow to extinguish conditioned fear responses and show enhanced startle when fear-primed. These proximate effects are manifestations of selection for risk avoidance in the sex which has primary responsibility for ensuring the protection and survival of dependent offspring. This enhanced fear system is simply another mechanism designed to ensure women stay alive for as long as possible.

References

Andreano, J. M., Dickerson, B. C., & Barrett, L. F. (2014). Sex differences in the persistence of the amygdala response to negative material. *Social Cognitive and Affective Neuroscience, 9*, 1388–1394.

Arrindell, W. A., & Buikhuisen, M. (1992). Dissimulation and the sex difference in self-assessed fears: A brief note. *Behavior Research and Therapy, 30*(3), 307–311.

Austad, S. N. (2006). Why women live longer than men: Sex differences in longevity. *Gender Medicine, 3*, 79–92.

Bekker, M. H. J., & van Mens-Verhulst, J. (2007). Anxiety disorders: Sex differences in prevalence, degree, and background, but gender-neutral treatment. *Gender Medicine, 4*, S178–S193.

Bjorklund, D. F., & Kipp, K. (1996). Parental investment theory and gender differences in the evolution of inhibition mechanisms. *Psychological Bulletin, 120*, 163–188.

Boyd, R., & Richerson, P. J. (1988). *Culture and the evolutionary process*. Chicago: University of Chicago Press.
Brønnum-Hansen, H., & Juel, K. (2001). Abstention from smoking extends life and compresses morbidity: A population based study of health expectancy among smokers and never smokers in Denmark. *Tobacco Control, 10*, 273–278.
Bullivant, S. B., Sellergren, S. A., Stern, K., Spencer, N. A., Jacob, S., Mennella, J. A., & McClintock, M. K. (2004). Women's sexual experience during the menstrual cycle: Identification of the sexual phase by noninvasive measurement of luteinizing hormone. *Journal of Sex Research, 41*, 82–93.
Campbell, A. (1999). Staying alive: Evolution, culture and intra-female aggression. *Behavioral and Brain Sciences, 22*, 203–252.
Campbell, A. (2008). Attachment, aggression and affiliation: The role of oxytocin in female social behaviour. *Biological Psychology, 77*, 1–10.
Campbell, A. (2013). *A mind of her own: The evolutionary psychology of women* (2nd ed.). Oxford: Oxford University Press.
Campbell, A. (2015). The evolutionary psychology of women's competition and aggression. In D. Buss (Ed.), *Handbook of evolutionary psychology* (2nd ed., pp. 684–703). Hoboken: Wiley.
Campbell, A., & Muncer, S. (1987). Models of anger and aggression in the social talk of women and men. *Journal for the Theory of Social Behaviour, I7*(4), 489–511.
Campbell, A., Muncer, S., & Bibel, D. (2001). Women and crime: An evolutionary approach. *Aggression and Violent Behaviour, 4*, 481–497.
Canli, T., Desmond, J. E., Zhao, Z., & Gabrieli, J. D. (2002). Sex differences in the neural basis of emotional memories. *Proceedings of the National Academy of Sciences, 99*(16), 10789–10794.
Caruso, S., Agnello, C., Malandrino, C., Lo Presti, L., Cicero, C., & Cianci, S. (2014). Do hormones influence women's sex? Sexual activity over the menstrual cycle. *The Journal of Sexual Medicine, 11*, 211–221.
Christiansen, D. M., & Hansen, M. (2015). Accounting for sex differences in PTSD: A multivariable mediation model. *European Journal of Psychotraumatology, 6*, 26068.
Closson, L. M. (2009). Status and gender differences in early adolescents' descriptions of popularity. *Social Development, 18*(2), 412–426.
Clutton-Brock, T. H., & Isvaran, K. (2007). Sex differences in ageing in natural populations of vertebrates. *Proceedings of the Royal Society B: Biological Sciences, 274*, 3097–3104.
Copping, L. T., & Campbell, A. (2015). The environment and life history strategies: Neighbourhood and individual-level models. *Evolution and Human Behavior, 36*, 182–190.
Copping, L. T., Campbell, A., & Muncer, S. (2013). Impulsivity, sensation seeking and reproductive behaviour: A life history perspective. *Personality and Individual Differences, 54*, 908–912.
Costa, P. T., Terracciano, A., & McCrae, R. R. (2001). Gender differences in personality traits across cultures: Robust and surprising findings. *Journal of Personality and Social Psychology, 81*(2), 322–331.
Cover, K. K., Maeng, L. Y., Lebrón-Milad, K., & Milad, M. R. (2014). Mechanisms of estradiol in fear circuitry: implications for sex differences in psychopathology. *Translational Psychiatry, 4*, e422–e422.
Cross, C. P., Copping, L. T., & Campbell, A. (2011). Sex differences in impulsivity: A meta-analysis. *Psychological Bulletin, 137*(1), 97–130.
Cross, C. P., Cyrenne, D.-L. M., & Brown, G. R. (2013). Sex differences in sensation-seeking: A meta-analysis. *Scientific Reports, 3*, 2486.
Day, H. L. L., & Stevenson, C. W. (2019). The neurobiological basis of sex differences in learned fear and its inhibition. *European Journal of Neuroscience, 52*, 1–21.
de Jonge, P., Roest, A. M., Lim, C. C. W., Florescu, S. E., Bromet, E. J., Stein, D. J., Harris, M., Nakov, V., Caldas-de-Almeida, J. M., Levinson, D., Al-Hamzawi, A. O., Haro, J. M., Viana, M. C., Borges, G., O'Neill, S., de Girolamo, G., Demyttenaere, K., Gureje, O., Iwata, N., Lee, S., Hu, C., Karam, A., Moskalewicz, J., Kovess-Masfety, V., NavarroMateu, F., Browne, M. O., Piazza, M., Posada-Villa, J., Torres, Y., ten Have, M. L., Kessler, R. C., & Scott, K. M. (2016).

Cross-national epidemiology of panic disorder and panic attacks in the world mental health surveys. *Depression and Anxiety, 33*, 1155–1177. https://doi.org/10.1002/da.22572.
Debiec, J. (2005). Peptides of love and fear: Vasopressin and oxytocin modulate the integration of information in the amygdala. *Bioessays, 27*, 869–873.
Del Giudice, M., & Ellis, B. J. (2016). Evolutionary foundations of developmental psychopathology. In D. Cicchetti (Ed.), *Developmental psychopathology. Vol. 2: Developmental neuroscience* (3rd ed., pp. 1–58). New York: Wiley.
Domes, G., Lischke, A., Berger, C., Grossmann, A., Hauenstein, K., Heinrichs, M., & Herpertz, S. C. (2010). Effects of intranasal oxytocin on emotional face processing in women. *Psychoneuroendocrinology, 35*, 83–93.
Drevenstedt, G. L., Crimmins, E. M., Vasunilashorn, S., & Finch, C. E. (2008). The rise and fall of excess male infant mortality. *Proceedings of the National Academy of Science U S A, 105*, 5016–5021.
Eder, D. (1985). The cycle of popularity: Interpersonal relations among female adolescents. *Sociology of Education, 58*, 154–165.
Eibach, R. P., & Mock, S. E. (2011). The vigilant parent: Parental role salience affects parents' risk perceptions, risk-aversion, and trust in strangers. *Journal of Experimental Social Psychology, 47*, 694–697.
Ellis, B. J., Figueredo, A. J., Brumbach, B. H., & Schlomer, G. L. (2009). Fundamental dimensions of environmental risk. The impact of harsh versus unpredictable environments on the evolution and development of life history strategies. *Human Nature, 20*, 204–268.
Fessler, D. M. T., Holbrook, C., Pollack, J. S., & Hahn-Holbrook, J. (2014). Stranger danger: Parenthood increases the envisioned bodily formidability. *Evolution and Human Behavior, 35*, 109–117.
Foerster, S., Franz, M., Murray, C. M., Gilby, I. C., Feldblum, J. T., Walker, K. K., & Pusey, A. E. (2016). Chimpanzee females queue but males compete for social status. *Scientific Reports, 6*, 35404.
Franklin, C. A. (2010). Physically forced, alcohol-induced and verbally coerced sexual victimization: Assessing risk factors among university women. *Journal of Criminal Justice, 38*, 149–159.
Gammie, S. C., Negron, A., Newman, S. M., & Rhodes, J. S. (2004). Corticotropin releasing factor inhibits maternal aggression in mice. *Behavioral Neuroscience, 118*, 805–814.
Grillon, C., Krimsky, M., Charney, D. R., Vytal, K., Ernst, M., & Cornwell, B. (2013). Oxytocin increases anxiety to unpredictable threat. *Molecular Psychiatry, 18*(9), 958–960.
Hahn-Holbrook, J., Holt-Lunstad, J., Holbrook, C., Coyne, S. M., & Lawson, E. T. (2011). Maternal defense: Breast feeding increases aggression by reducing stress. *Psychological Science, 22*(10), 1288–1295.
Hard, E., & Hansen, S. (1985). Reduced fearfulness in the lactating rat. *Physiology and Behavior, 35*, 641–643.
Haring, R., Volzke, H., Steveling, A., Krebs, A., Felix, S. B., Schofl, C., Dorr, M., Nauck, M., & Wallaschofski, H. (2010). Low serum testosterone levels are associated with increased risk of mortality in a population-based cohort of men aged 20–79. *European Heart Journal, 31*, 1494–1501.
Herold, E. S., & Mewhinney, D. M. K. (1993). Gender differences in casual sex and aids prevention: A survey of dating bars. *Journal of Sex Research, 30*, 36–42.
Hofstede, G. (1991). *Cultures and organizations: Software of the mind*. London: McGraw-Hill.
Hsu, K., & Iwamoto, D. K. (2014). Testing for measurement invariance in the conformity to Masculine Norms-46 across White and Asian American college men: Development and validity of the CMNI-29. *Psychology of Men and Masculinity, 15*, 397–406.
James, J., Ellis, B. J., Schlomer, G. L., & Garber, J. (2012). Sex specific pathways to early puberty, sexual debut and sexual risk taking: Test of an integrated evolutionary-developmental model. *Developmental Psychology, 48*, 687–702.
Kruger, D. J., & Nesse, R. M. (2006). An evolutionary life-history framework for understanding sex differences in human mortality rates. *Human Nature, 17*, 74–97.

Leckman, J. F., Mayes, L. C., Feldman, R., Evans, D., King, R. A., & Cohen, D. J. (1999). Early parental preoccupations and behaviors and their possible relationship to the symptoms of obsessive compulsive disorder. *Acta Psychiatrica Scandinavica, 100*, 1–26.

LeDoux, J. (2012). Rethinking the emotional brain. *Neuron, 73*, 653–676.

Lee, S. Y., Kesebir, S., & Pillutla, M. M. (2016). Gender differences in response to competition with same-gender coworkers: A relational perspective. *Journal of Personality and Social Psychology, 110*(6), 869–886.

Lemaire, J. (2013). Why do females live longer than males? *North American Actuarial Journal, 6*, 21–37.

Lindahl-Jacobsen, R., Hanson, H. A., Oksuzyan, A., Mineau, G. P., Christensen, K., & Smith, K. R. (2013). The male–female health-survival paradox and sex differences in cohort life expectancy in Utah, Denmark, and Sweden 1850–1910. *Annals of Epidemiology, 23*, 161–166.

Lischke, A., Gamer, M., Berger, C., Grossmann, A., Hauenstein, K., Heinrichs, M., Herpertz, S. C., & Domes, G. (2012). Oxytocin increases amygdala reactivity to threatening scenes in females. *Psychoneuroendocrinology, 37*(9), 1431–1438.

Luy, M. (2003). Causes of male excess mortality: Insights from cloistered populations. *Population and Development Review, 29*, 647–676.

McDonald, M. M., Navarette, C. D., & Van Vugt, M. (2012). Evolution and the psychology of intergroup conflict: The male warrior hypothesis. *Philosophical Transactions: Biological Sciences, 367*(1589), 670–679.

Merz, C. J., Kiner, V. L., & Wolf, O. T. (2018). Let's talk about sex . . . differences in human fear conditioning. *Current Opinion in Behavioral Sciences, 23*, 7–12.

Mesoudi, A. (2016). Cultural evolution: Integrating psychology, evolution and culture. *Current Opinion in Psychology, 7*, 17–22.

Moriguchi, Y., Touroutoglou, A., Dickerson, B. C., & Barrett, L. F. (2014). Sex differences in the neural correlates of affective experience. *Social Cognitive and Affective Neuroscience, 9*, 591–600.

Morris, J. A., & Harrison, L. M. (2009). Hypothesis: Increased male mortality caused by infection is due to a decrease in heterozygous loci as a result of a single X chromosome. *Medical Hypotheses, 72*, 322–324.

Olff, M., Langeland, W., Draijer, N., & Gersons, B. P. R. (2007). Gender differences in posttraumatic stress disorder. *Psychological Bulletin, 133*(2), 183–204.

Pedersen, C. A. (2004). Biological aspects of social bonding and the roots of human violence. *Annals of the New York Academy of Sciences, 1036*, 106–127.

Reidy, D. E., Brookmeyer, K. A., Gentile, B., Berke, D. S., & Zeichner, A. (2016). Gender role discrepancy stress, high-risk sexual behavior, and sexually transmitted disease. *Archives of Sexual Behavior, 45*(2), 459–465.

Stockley, P., & Campbell, A. (2013). Female competition and aggression: Interdisciplinary perspectives. *Philosophical Transactions of the Royal Society B, 368*, 20130073.

Sutton, R. M., Robinson, B., & Farrall, S. D. (2011). Gender, fear of crime, and self-presentation: An experimental investigation. *Psychology, Crime and Law, 17*(5), 421–433.

Taylor, S. E., Klein, L. C., Lewis, B. P., Gruenewald, T. L., Gurung, R. A. R., & Updegraff, J. A. (2000). Biobehavioral responses to stress in females: Tend-and-befriend, not fight-or-flight. *Psychological Review, 107*(3), 411–429.

United Nations. (2011). *Sex differentials in childhood mortality*. New York: United Nations, Department of Economic and Social Affairs, Population Division.

Vandello, J. A., & Bosson, J. K. (2013). Hard won and easily lost: A review and synthesis of theory and research on precarious manhood. *Psychology of Men and Masculinity, 14*(2), 101–113.

Verlasco, E. R., Florido, A., Milad, M. R., & Andero, R. (2019). Sex differences in fear extinction. *Neuroscience and Biobehavioral Reviews, 103*, 81–108.

Wager, T. D., Phan, K. L., Liberzon, I., & Taylor, S. F. (2003). Valence, gender, and lateralization of functional brain anatomy in emotion: A meta-analysis of findings from neuroimaging. *NeuroImage, 19*, 513–531.

Waldron, I. (1983). Sex differences in human mortality: The role of genetic factors. *Social Sciences and Medicine, 17*, 321–333.
Wood, W., & Eagly, A. H. (2002). A cross-cultural analysis of the behaviour of women and men: Implications for the origins of sex differences. *Psychological Bulletin, 128*, 699–727.
Wood, W., & Eagly, A. H. (2012). Biosocial construction of sex differences and similarities in behaviour. *Advances in Experimental Social Psychology, 46*, 55–123.
Xing, D., Nozell, S., Chen, Y. F., Hage, F., & Oparil, S. (2009). Estrogen and mechanisms of vascular protection. *Arteriosclerosis, Thrombosis, and Vascular Biology, 29*, 289–295.
Young, K. D., Bellgowan, P. S., Bodurka, J., & Drevets, W. C. (2013). Functional neuroimaging of sex differences in autobiographical memory recall. *Human Brain Mapping, 34*(12), 3320–3332.
Zarulli, V., Barthold Jones, J. A., Oksuzyan, A., Lindahl-Jacobsen, R., Christensen, K., & Vaupel, J. W. (2018). Women live longer than men even during severe famines and epidemics. *Proceedings of the National Academy of Sciences of the U S A, 115*(4), E832–E840.

Appendix A

Experimental techniques used in fear studies in relation to proposed components.

Technique	Method	Resulting measure	Component relevance
CO_2 inhalation	CO_2 concentration (normally 0.04%) is increased to between 0 and 35%. Via respiratory mask with head strap, participant takes fixed number and length of breaths (e.g. 12 consecutive 20s inhalations). Panicogenic	Sympathetic nervous system reactivity (e.g. skin conductance responses, frontalis EMG, heart rate)	Sensitivity
		Subjective pre- and post- ratings (e.g. Anxiety Sensitivity Index, Subjective Units of Distress, State-Trait Anxiety)	Awareness (self-report)
Dot probe	Two stimuli (one of which is threatening) appear simultaneously at different locations on a monitor (e.g. right and left). After a short exposure time both stimuli disappear, a probe appears at one of the two locations and participants are asked to indicate as quickly as possible its spatial location. A bias toward threat is inferred when the reaction time on congruent trials (the probe appears at the same location as the threatening stimulus) is faster than incongruent trails (the probe appears at the location of the neutral stimulus)	A bias index is computed by subtracting congruent reaction times from incongruent trials so that a positive value indicates bias toward threat (attentional vigilance) and a negative value a bias away from threat (attentional avoidance). An Orienting index measuring facilitated orientation calculated by subtracting the mean reaction times of congruent trials from mean reaction times of neutral trials. A disengaging index measuring delayed disengagement calculated by subtracting mean reaction times of neutral trials from mean reaction times of incongruent trials	Sensitivity

(continued)

A. Campbell et al., *Sex Differences in Fear Response*, SpringerBriefs in Anthropology, https://doi.org/10.1007/978-3-030-65280-7

Technique	Method	Resulting measure	Component relevance
Emotional Stroop	Participants asked to read the ink colour of printed words and ignore meaning. Threatening words ("die") are compared to neutral ("sum"), or pleasant words ("joy"). Colour naming is slower when the distracter word is threatening suggesting difficulty in filtering out threatening information	Event related potentials (via ECG) or reaction times	Sensitivity
Fear conditioning and extinction	Classical conditioning of fear pairs a neutral stimulus such as a light (conditioned stimulus, CS^+) with an aversive stimulus such as a mild shock (unconditioned stimulus, UCS) until a fear response occurs to the CS^+ alone. A second CS (CS^-) is not paired with a UCS. Following this acquisition procedure, extinction trials, extinction memory trials and reinstatement trials may be introduced	The difference in response to CS^+ and CS^- (usually SCR) is taken as an index of fear conditioning	Time course
Neuroimaging (Awareness)	The anterior insulate cortex is often co-activated with the anterior cingulate cortex. The anterior insulate cortex appears to be more strongly involved in interoception while the anterior cingulate cortex mediates motor response. A common feature of tasks eliciting the anterior insulate cortex but not the anterior cingulate cortex activation is awareness	Selective regional activation of anterior insulate cortex without anterior cingulate cortex involvement	Awareness
Neuroimaging (Salience)	Identification of salience network is based on activation during key tasks. Tracks visceral information, emotions and pain associated with bodily states, guides motor response. Shows strong activa-tion in response to oddball (novel or salient) stimuli across multiple modalities. Unique to these cortical regions, the anterior insulate	Selective regional activation in anterior insulate cortex and anterior cingulate cortex to the presence of emotional, pain-relevant and 'oddball' stimuli	Salience

(continued)

Technique	Method	Resulting measure	Component relevance
	cortex and anterior cingulate cortex contain von Economo neurons, thought to relay information to frontal and temporal regions to assess complex situations such as threat		
Neuroimaging studies (Strength)	Emotional images, usually facial but sometimes scenes, are presented in a scanner. Participants may be instructed to attend to irrelevant aspects of the image to maintain attention. This is usually done using fMRI technology and less often other scanning and tracing techniques	Selected regional activation in the viewers' brain (regions of interest, sometimes supplemented with whole brain measures) is recorded and compared between critical (emotional) and neutral (non-emotional) trials	Strength
Potentiated startle reflex	Sudden onset acoustic tone presented during fear-related conditions (threatening film or anticipation of electric shock)	Difference in eye muscle contraction strength between neutral and potentiated conditions	Duration
Risky decision making	Widely used tasks include (1) Iowa Gambling Task: Participants select a series of cards from one of four decks which vary in the size of rewards, penalties and the frequency of incurring them (see text). (2) The Balloon Analogue Risk Task (BART) Participants inflate a computer-simulated balloon for prize points which can be lost if the balloon bursts (see text). (3) Various gambling games that manipulate odds, wins and losses	Number or percentage of risky choices selected	Salience
Self-reported fear	Respondents (1) complete a standardized psychometric inventory of rating their fear of a range of stimuli (2) rate the frequency of experiencing fear within a specified time window or (3) rate the typical intensity of a fear episode	Standardized measure; or a sum or average of intensity or frequency	Awareness Time course (memory dependence)
Startle reflex	Sudden onset acoustic tone presented during neutral film viewing or stimulus conditions	Contraction of eye muscles (orbicularis oculi)	Sensitivity

(continued)

Technique	Method	Resulting measure	Component relevance
Sustained fear	When an initial threat persists in time, amygdala activation is superseded by activation of the bed nucleus of the atria terminalis which maintains anticipatory watchfulness ('sustained fear')	Startle magnitude is compared between predictable shock (signalled by a tone) and unpredictable shock (associated with different virtual reality contexts) see Alvarez, Chen, Bodurka, Kaplana, and Grillon (2011)	Time Course

Appendix B

Summary of regions and circuitry functionally implicated in fear.

Amygdala	Orienting responses to motivationally relevant exteroceptive stimuli which are novel, unusual, or associated with uncertain consequences (prediction errors)
Lateral amygdala	• Receives sensory auditory, visual and somatosensory inputs about unconditioned and conditioned fear stimuli and cortical inputs from dorsolateral anterior cingulate cortex, orbital and medial prefrontal cortex • Responsible for condition-induced plasticity (altered magnitude of neuronal response to Conditioned Stimuli) that drives conditioning
Basolateral amygdala	• Receives incoming input from the hippocampus • Contains 3 populations of neurons: 1. 'Persistent neurons' that encode the maintenance of Conditioned Stimuli-Unconditioned Stimuli associations after learning 2. 'Fear neurons' that encode fear expression 3. 'Extinction neurons' that encode the formation of extinction memories
Central amygdala	• Coordinates output • Projections to the periaqueductal grey drive fear behaviour • Projections to hypothalamus, midbrain, pons, and medulla control activation of sympathetic nervous system and hypothalamic-pituitary-adrenal axis • Can alter vigilance and attention to exteroceptive stimuli by modulating response thresholds of sensory cortical neurons
Anterior insula	• Monitors interoceptive cues • Awareness of bodily sensations and affect (emotion, hedonic tone, arousal) • Activated during tasks involving subjective awareness of positive and negative feelings • Implicated in empathy (co-affect) • Involved in transient detection of salient stimuli and the initiation of attentional control • With dorsolateral anterior cingulate cortex and amygdala, constitutes

(continued)

A. Campbell et al., *Sex Differences in Fear Response*, SpringerBriefs in Anthropology, https://doi.org/10.1007/978-3-030-65280-7

	a salience network that identifies the most relevant interoceptive and exteroceptive stimuli in order to guide behaviour
Dorsal anterior cingulate cortex[a]	• Dorsolateral anterior cingulate cortex (fear enhancement) plays an opposite role to the ventromedial prefrontal cortex (fear reduction) in the regulation of fear behaviour • Involved in the expression of behavioural and autonomic fear responses • Activated during fear conditioning. Also active to Unconditioned stimuli and high intensity stimuli • During fear expression, functional coupling with ventromedial prefrontal cortex, amygdala, and hippocampus • Sustains top-down attention control on salient stimuli guiding response selection and motor control • With the anterior insulae and amygdala, constitutes a salience network that identifies the most relevant interoceptive and exteroceptive stimuli in order to guide behaviour
Hippocampus	• Encodes contextual features associated with fear expression • Activated jointly with the ventromedial prefrontal cortex during extinction recall
Hypothalamus	• Manages hypothalamic-pituitary-adrenal axis and autonomic fear responses
Orbitofrontal cortex[b]	• Emotional recognition and subjective experience of emotion • Control of negative emotions via attention deployment, re-appraisal and response suppression • Aggression and anger • Registers prediction error signals (unexpected events) • Flexibly tracks stimulus-outcome associations required for reversal or extinction learning • Computes net economic value of choices to inform decision making • Contributes to 'cognitive map' that defines the current task space by integrating current perception, past experiences and prediction error signal to simulate potential future outcomes that guide decision-making
Periaqueductal grey	• Involved in detection of internal physiological signals that trigger defensive behaviour • Output to motor areas directs behavioural fear reactions • Regulates autonomic substrates that allow for behavioural fear reactions • Modulates pain
Ventromedial prefrontal cortex[c] vmPFC	• Ventromedial prefrontal cortex (fear reduction) plays an opposite role to dorsolateral anterior cingulate cortex (fear enhancement) in the regulation of fear behaviour • Activated during fear extinction. Deactivated during conditioning • Modulates amygdala activity • During extinction, functional coupling with amygdala and hippocampus • Associated with consolidation and retrieval of extinction fear memory. Increased activation during extinction recall

[a]Homologous to rodent prelimbic prefrontal cortex
[b]Located within the ventromedial prefrontal cortex. One of the least understood regions, see Stalnaker, Cooch, and Schoenbaum (2015)
[c]Homologous to rodent infralimbic prefrontal cortex

Appendix C: Cohen's d

Cohen's d takes the difference in means between two groups (here, men and women) and divides it by the pooled standard deviation. This means that the difference between men and women is expressed as a proportion of the variability that exists in the population as a whole. The table below 'translates' a range of values of Cohen's d into statements about the relative scores of men and women on any measure. Note that this table assumes that men are the reference category, so for measures on which women score higher than men, a negative sign would be placed before the d value and the size of the effect would be interpreted the same way.

Cohen's d	Rule of thumb[a]	Average man scores higher than...[b, c]	Overlap between men's and women's distributions is...[c] (%)	If I know someone's score, I can correctly guess their sex...[b, d]
0	No effect	50% of women	100	50% of the time
0.2	Small effect	58% of women	92	54% of the time
0.5	Medium effect	69% of women	80	60% of the time
0.8	Large effect	79% of women	69	66% of the time
1.0	Very large effect	84% of women	62	69% of the time

[a]This rule of thumb was put forward by Cohen himself, who noted that interpreting effects in this way is "an operation fraught with many dangers" (Cohen, 1977)
[b]For more information, see Coe (2002)
[c]For more information including an interactive visualisation, see Magnusson (n.d.) rpsychologist.com/cohend/
[d]Assumes that the population of interest consists of 50% men and 50% women

A. Campbell et al., *Sex Differences in Fear Response*, SpringerBriefs in Anthropology, https://doi.org/10.1007/978-3-030-65280-7

The manufacturer's authorised representative in the EU is Springer Nature Customer Service Centre GmbH, Europaplatz 3, 69115 Heidelberg, Germany. If you have any concerns regarding our products, please contact ProductSafety@springernature.com

Printed and bound by CPI Group (UK) Ltd, Croydon, CR0 4YY

30/09/2026

02225708-0005